The Competitive Economics of Nuclear and Coal Power

The Competitive Economics of Nuclear and Coal Power

Richard Hellman
Caroline J.C. Hellman
University of Rhode Island

LexingtonBooks
D.C. Heath and Company
Lexington, Massachusetts
Toronto

Library of Congress Cataloging in Publication Data

Hellman, Richard, 1913–
  The competitive economics of nuclear and coal power.

  Bibliography: p.
  Includes index.
  1. Electric utilities—United States—Costs—Case studies. 2. Atomic
power-plants—United States—Costs—Case studies. 3. Coal-fired power
plants—United States—Costs—Case studies. I. Hellman, Caroline J.C.
II. Title.
HD9685.U5H39   1982   338.2'3   82–47500
ISBN 0–669–05533–6

Copyright © 1983 by D.C. Heath and Company

Published simultaneously in Canada

Printed in the United States of America

International Standard Book Number: 0–669–05533–6

Library of Congress Catalog Card Number: 82–47500

Contents

Contents

Tables

Acknowledgments

Numerous people contributed to the writing of this book, and we are grateful to them all. Those who commented on various parts of the manuscript and provided general encouragement and support include: Theron Raines, of Raines and Raines, New York City; Mason Wilson, professor of engineering and director of the University Energy Center at the University of Rhode Island (URI); Nat Sage, coordinator of research at URI; Dilip Datta, professor of mathematics at URI; Glen Ramsay, Art Mead, and Joel Dirlam, professors in the Economics Department at URI; and Doug Jones, director of the National Regulatory Research Institute at The Ohio State University.

Part of the early research was funded by the Department of Energy. We appreciated the assistance of the staffs at the Nuclear Regulatory Commission, Department of Energy (including the Federal Energy Regulatory Commission), Internal Revenue Service, Exxon's Research and Engineering Company, and the Mechanical Engineering Division of American Electric Power.

Special thanks go to Mark Farber at Temple, Barker and Sloane, Inc., Lexington, Massachusetts, who commented on the entire manuscript and tirelessly debated various points; to Violet Hellman for her assistance with the manuscript and invaluable support; to Denise Foley who brought unusual skill and cheerfulness to the task of typing the manuscript; and to the staff of Lexington Books for their patience and help.

Introduction

The objective of this book is to analyze the competitive economics of nuclear and coal power plants newly ordered for initial commercial operation in 1985-1995. The focus is on bottom-line costs and on identifying and analyzing the engineering and economic factors that a valid study should address.

The development of nuclear-power costs is singularly problematic. In principle, estimates for the various cost components (for example, capital, fuel, and operating and maintenance) should be obtained primarily from the architect/engineers (A/E) and manufacturers of nuclear power plants. Such estimates, to date, have greatly underestimated realized costs. For example, Philadelphia Electric Company and/or Bechtel Corporation (the A/E) revised the capital-cost estimates for the Limerick nuclear power plant eleven times between May 1968 and October 1980. Total costs rose from $326 million to $4.12 billion. The estimates as of March 26, 1982, had further increased to a range of $4.23 to $6.6 billion (including noncompany estimates).[1]

An alternative to costs developed by A/Es and manufacturers is regression analysis. Extrapolating from a conglomeration of past plant costs to future estimates is especially speculative and uncertain in an industry marked by engineering and economic unpredictability.[2]

A third approach has been used here: begin with the A/E's costs (available only to various government and industry sources) but adjust the estimate, to the extent feasible, for unrealistic assumptions and omissions. The analysis is based on four of the best case studies of nuclear and coal economics whose sponsors are neutral, if not sympathetic, toward nuclear power. The studies include:

The Atomic Energy Commission (AEC) studies of 1974-1975

The Energy Research and Development Administration (ERDA) study of December 1976

The Nuclear Regulatory Commission (NRC) environmental-impact statement of 1979 on the proposed nuclear power plant for Charlestown, Rhode Island

The Exxon study of 1977-1979, unpublished but made available to one of the authors of this book

The latter two studies, being more recent and sounder methodologically than the other two, are the more important.

Fifteen engineering/economic factors are identified and analyzed and, where feasible, quantified and incorporated into the four case studies. They include:

Technological sufflation, that is, capital-cost inflation independent of the general inflation rate and financing costs

Construction times

Economic life of units

Capacity factor

Senescence, that is, the degradation of capacity factors with age of plant

Btu per pound of coal

Heat rate, that is, efficiency of coal combustion

Fuel prices

Operating and maintenance (O&M) costs

The "Yoyo Factor," that is, predictability of unit availability

Replacement power

The economics and accounting procedures for major postconstruction repairs for nuclear plants

The economics of low-sulfur coal

Waste disposal for nuclear plants

Decommissioning for nuclear plants

The adjusted results of the four case studies indicate an unexpectedly large gap between the total average costs (mills per kwh) of nuclear and coal power; after adjustments, coal ranges from 22 percent to 50 percent cheaper than nuclear (without including several factors for nuclear that could not be quantified). How can these results be reconciled with the widely held belief that nuclear power is a cheap form of energy? The answer, and the key to the economic failure of nuclear power, is that a very wide gap exists between the design and the actual performance of nuclear power units.

The design for virtually all nuclear units operating as of 1982 in the United States specifies two key performance standards, namely,

An 80-percent capacity factor (or better)

A level of control (predictability) of availability necessary for baseload operation

During the past twenty-two years of commercial operation of nuclear power plants, the annual average nameplate capacity factor has run between 53 percent and 60 percent. The averages since the Three Mile Island accident have been close to 55 percent. Nuclear units have also experienced a fairly high degree of uncontrolled availability, with units frequently restricting or shutting down operation unexpectedly at primary and secondary peaks.

The economic problems of nuclear power would not be solved by increasing capacity factors, that is, the number of kilowatt hours (kwh) over which total costs are averaged. The problems also result from substantially higher total costs than those accounted for in the design economics. These additional costs include higher O&M expenses, major postconstruction repairs, replacement power, and uncertain but burgeoning waste-disposal and decommissioning costs.

Four questions should be asked concerning the design/performance gap of nuclear power:

To what can the design/performance gap be attributed?

Who should assume responsibility for the performance of nuclear power plants?

While nuclear plants have not operated well to date, has performance improved over time, that is, has the industry exhibited learning?

Do the answers to the preceding three questions provide insight into how the design/performance gap might be narrowed?

The first three questions are addressed in part I, which discusses two primary causes for nuclear's design/performance gap: design-technological risks and human factors. In response to the second question (on responsibility), the Bechtel Theorem is introduced, which documents the fact that the nuclear power plant A/E does not guarantee performance and which quantifies the costs of providing such a guarantee. Evidence for learning—the third question—is examined from a new perspective. Avenues for narrowing the design/performance gap (question 4) follow from the discussion of the first three questions.

Part II describes the methodology used to adjust the case studies, including the assessment of the engineering/economic factors. The numerical results, before and after adjustments, for the case studies are presented in part III, followed by a brief summary and conclusions.

The appendix analyzes an interesting policy issue for the nuclear power industry—whether certain tax and rate-making procedures have encouraged premature commerciality of nuclear power plants. Premature commerciality results in financial benefits for electric utilities and possible safety hazards. The confused state of much data on the nuclear power industry is also addressed.

The policy implications of this study should be noted. The political, ethical, or environmental arguments for or against nuclear or coal power are not addressed here. The objective is to provide the most realistic projections of nuclear and coal generating costs. The decision to use nuclear or coal power must assess realistically the prospects for closing the design/performance gap and balance the resulting economics against the numerous noneconomic decision criteria.

Notes

1. Pennsylvania Public Utility Commission, *Limerick Nuclear Generating Station Investigation,* Initial Decision of Administrative Law Judge, I–80100341, March 26, 1982, p. 71.

2. Regression analysis is discussed by two adversary experts in the Limerick case, Ibid., p. 82 ff.

Part I
The Design/Performance Gap
for Nuclear Power

1 Introduction to Part I

We [the nuclear industry] got going too fast. We scaled-up nuclear reactors too fast, handed them out to the utilities too fast, and the technology wasn't quite ready, and the utilities weren't quite ready. With nuclear submarines, Admiral Rickover set up an organization for the Navy that reflected the demands of the technology. The utilities didn't do that. It's just as if you had handed the airlines over to the railroads to run.[1]

—Victor Gilinsky, Commissioner
Nuclear Regulatory Commission

Overview

The commercial performance of U.S. nuclear power units has fallen far short of nuclear power's promise.[2] Part I discusses the primary causes for the design/performance gap and examines whether the gap appears to be narrowing. The origins of the gap provide insight into how it might be closed.

Chapter 2 addresses the design-technological risks of nuclear power. Such risks include the scale (that is, MW size) of units; corrosion in the radioactive and conventional circuits of a unit; metallurgy (especially welding); and tubing, piping, valves, pumps, and controls. Human factors in the design, building, operating, and decommissioning of nuclear power plants are discussed in chapter 3. An issue related to design-technological risks and to human factors—premature commerciality of nuclear units—is analyzed in the appendix. The appendix also examines the confused state of much data on the nuclear power industry.

Chapter 4 analyzes learning (that is, closing the design/performance gap) in the U.S. nuclear power industry to date. Four performance measures are compared across younger and older units, to wit:

Number of days from a unit's initial criticality to commerciality

Capacity factor for the first twelve months of commercial operation

1979 annual capacity factor

Cumulative-lifetime-capacity factor through December 1979

3

Before proceeding, a recent court case is discussed that encapsulates three salient aspects of the design/performance gap, namely,

Recognition of the gap by industry

Financial responsibility for the gap

The economics of closing the gap

The Bechtel Theorem

The Bechtel Theorem emerges from proceedings before the U.S. District Court of Oregon. At issue were the unexpected increases in costs of the Trojan nuclear power plant; the lawsuit of its principal owner—Portland General Electric Company (PGE)—to recover these costs from Bechtel Corporation, the architect/engineer (A/E); and Bechtel's grounds for counterclaims.

The Trojan nuclear power unit was ordered in 1968 at an announced estimated cost of $228 million, with Bechtel as the architect/engineer. The unit was declared commercial in May 1976. It immediately encountered problems, as evidenced by its annual capacity factors (nameplate):

Year	Trojan's Capacity Factor
1976	19.7%
1977	61.0
1978	15.6
1979	49.4
1980	56.9
1981	60.3
Lifetime, through 1981	46.0

Source: Adapted from Nuclear Regulatory Commission, *Operating Units Status Report,* monthly (Washington, D.C.).

In February 1979 PGE sued Bechtel for damages arising out of its difficulties with Trojan, direct and consequential, including the cost of replacement power during shutdowns. In its reply, Bechtel stated that, at the time Trojan was ordered, the risks of nuclear power were extraordinary, unknown, and unquantifiable. Therefore, Bechtel stated, its fee as A/E did not include any item for such risks, consequential and otherwise. The language is worth quoting:

As of 1968 the risk of consequential damages arising from the unplanned shutdown of an electric power generating plant and the resulting needs of the owner of the plant to purchase replacement power was extraordinarily variable, unpredictable and uncontrollable. In the electric power generation industry at that time there was a long standing custom and usage, recognized and accepted by all concerned parties, that architect-engineers for electric power generating plants were not to be liable for any form of consequential damages, including loss of profit and loss of use.[3]

The preceding is most relevant to nuclear power plants. For fossil-fuel plants, in general, costs, construction times, and operating results have been much more predictable and in line with design (see part II for data).

Bechtel quoted language in its contract with PGE specifying its immunity to such damages and asked the court, if PGE nevertheless were upheld, to award Bechtel compensation for assuming the extra risks:

> Should the limitation of liability provision . . . be held unenforceable, then Bechtel would be entitled to an adjustment in its compensation to reasonably reflect the value of an acceptance of the risk of liability for consequential damages . . . . In the event that Bechtel is held to have assumed the risk of consequential damages, despite its express understanding and contractual provisions to the contrary, then Bechtel is entitled to the reasonable value of the imposition upon it of this risk exposure at plaintiffs' instance and request . . . . Bechtel estimates the reasonable value of an acceptance of this risk exposure to be substantially in excess of $100,000,000. . . . This adjustment is necessary, in the event plaintiffs' claims for consequential damages are found valid, in order to prevent the unjust enrichment of plaintiffs by reason of their obtaining the services of Bechtel and the acceptance of risk exposure by Bechtel, at a price substantially below the reasonable value of Bechtel's services and acceptance of risk.[4]

Bechtel's assignment of "substantially" more than $100 million as the value of assuming the extra risks of the nuclear power plant compares with the originally announced total cost of $228 million, and with the $452 million actual cost (not including interest during construction) reported to the Federal Power Commission by the company.

The court, in its order of June 4, 1980, ruled that the agreement between Bechtel and PGE released Bechtel from consequential damages. It noted, for example, that Bechtel, being "afraid of nuclear spill," was careful to exclude consequential damages from the contract.

In a further decision of May 1981, the court entered an agreement between Bechtel and PGE resolving the entire suit. The court, however, ordered the substance of the settlement to be kept secret by the parties (*Wall Street Journal,* March 23, 1981).

The significance of the Bechtel PGE case cannot be overestimated. The Bechtel Theorem states that

> Consequences of unplanned nuclear-plant shutdowns are "extraordinarily variable, unpredictable, and uncontrollable" and that these

risks are "recognized and accepted by all concerned parties" in the electric-power-generation industry.

Responsibility for the risks of nuclear power is not included in the A/E's costs for a plant.

Assumption of nuclear power's risks would have, in the PGE case, increased the originally announced plant cost by at least 50 percent.

In short, the Bechtel Theorem constituted, some thirteen years (1968) after the industry's first nuclear plant was ordered and eight years after the first plant went commercial, a blank check for nuclear power costs. The record since 1968 has not invalidated the theorem.

Notes

1. *Life,* May 1982, p. 40. Reprinted with permission.
2. The performance of European nuclear-power units is discussed in part I as well.
3. Civil No. 79–103: "Answer, Counterclaim, and Demand for Jury Trial of Defendants Bechtel Corporation and Bechtel Power Corporation."
4. Ibid.

2 Design-Technological Risks of Nuclear Power

Comparative Nuclear and Fossil-Fuel Scaling

The design of nuclear power plants has involved a technological risk that has received little mention in the literature of nuclear-power economics, namely, the scaling up of megawatt capacity by jumps so big as to create design and operating problems for the reactor as well as for the conventional circuits. The scale of the conventional circuits has generally exceeded the physical dimensions already operating in fossil-fuel plants.

The Nuclear Division of Siemens in Erlangen (Germany) attributes the more frequent casualties in the conventional (as opposed to the reactor) circuit of nuclear power plants to scaling problems.[1] For the same MW capacity, the conventional components of a nuclear power plant (that is, turbogenerators, steam generators, piping, valves, and pumps) must be much bigger physically than those same components in a fossil-fuel plant. For example, if the largest fossil-fuel unit in operation were 1,300-MW and a 1,300-MW nuclear unit were ordered, the conventional part of the nuclear unit would exceed in physical dimensions the corresponding parts in the fossil-fuel unit. Thus the scale projections would exceed existing experience.

The history of scaling jumps is presented in table 2–1. The implications for the reactor and conventional circuits are discussed separately in the sections that follow.

Reactor Circuit

Nuclear power's first generation comprised three units of 185 to 285 MW, ordered in 1955–1956, and made commercial in 1960–1962. During the entire period 1962–1967, when orders rose from 600 to 1,170 MW, the largest nuclear unit in operation was 285 MW. Thus unit scale quadrupled without operating experience with an intermediate-size reactor.

Conventional Circuit

The conventional part of the first three 285-MW units was, when ordered, well within the operating experience of fossil-fuel units. However, the

7

Table 2–1

Size Escalation of Nuclear Power Units Awarded or Announced Compared with Largest Commercial Nuclear and Fossil-Fuel Units

| Year | Nuclear Unit Awarded/Announced[a] | | | Largest Commercial Unit (MW)[b] | |
	Name	MW	Nuclear	Fossil Fuel
1955–56	Dresden-1 } Indian Pt.-1 } Yankee Rowe }	185–285	—	c
1960	—	—	—	496 (Breed)
1962	Haddam Neck	600	285 (Ind. Pt.-1)	496
1963	Oyster Creek	670	285	496
1965	Dresden-2	828	285	550 (Col.-B)
1965	Indian Pt.-2	1,013	285	550
1966	Browns Ferry-1	1,152	285	550
1966	Salem	1,170	285	550
1967	—	—	285	950 (Bu. Rn.)
1968	Sequoyah-1&2	1,220	600 (Haddam)	950
1968	Fermi	1,290	600	950
1969	—	—	642 (Nine Mi. Pt.)	950
1970	—	—	642	1,150 (Par.-B)
1971	—	—	769 (Rob.-2)	1,150
1972	—	—	848 (Surry-1)	1,150
1973	6 units	1,300	1,098 (Zion-1)	1,300 (Cu. & Am.)
1974	10 units	1,300	1,152 (Br. Fy.-1 & Pch. Bot.)	1,300

Source: Announcements and awards from *Annual Reviews* of AEC, and FPC's annual *Statistics of Steam Electric Plants*. Capacities and commercial dates from NRC monthly *Status Report* and FPC *Statistics of Steam Electric Plants*.

[a]The year announced and awarded are approximately the same. Construction generally started one to two years after announcement/award. The distinction between announcement and award is not often clear in the sources. The announced and nameplate MW usually do not differ by more than 10–50 MW.

[b]The unit names abbreviated in the table are: Indian Point-1, Haddam Neck, Nine Mile Point, Robinson-2, Surry-1, Zion-1, Browns Ferry-1, Peach Bottom, Breed, Colbert-B, Bull Run, Paradise-B, Cumberland, and Amos.

[c]The 285-MW nuclear unit was well within the scale of operating fossil-fuel units.

600-MW Haddam Neck unit, ordered in 1962, exceeded the largest fossil-fuel unit then operating. When the 828-MW Dresden-2 unit was ordered in 1965, the 550-MW Colbert-B, which became commercial the same year, was the largest fossil-fuel unit in commercial operation. Subsequent extensions of commercially operating fossil-fuel sizes included Bull Run (950 MW) in 1967, Paradise (1,150 MW) in 1970, and Cumberland (1,300 MW) in 1973. In 1973 six nuclear units of 1300 MW were ordered, followed by another ten of this size in the first nine months of 1974. Two facts emerge from this scaling sequence:

The nuclear units ordered, after the first group of three, always substantially exceeded the largest sizes of fossil-fuel units already operating.

The scaling risk was compounded by the need for physically larger components than for the same fossil-fuel size.

Source of the Scaling Risk

Three of the factors causing the physical dimensions and scaling of nuclear power units to exceed those of fossil-fuel units are steam temperature, steam pressure per square inch (psi), and turbine revolutions per minute (rpm). Table 2–2, in which coal and nuclear units have been matched to the extent feasible by age and size,[2] indicates the following:

Temperature: Nuclear steam temperature (abut 500° to 575 °F) is slightly over half that of fossil fuel (1,000 °F or slightly higher).

Pressure: Most psis are in the 700 to 950 range for nuclear, but for fossil fuel they are mainly at 3,500.

RPM: Nuclear turbines seem to be standardized at 1,800 rpm, fossil fuel at 3,600.

The preceding three characteristics explain the greater volumes of fluids and steam that must be moved through the pipes, valves, pumps, and steam generator of the conventional circuit of nuclear power plants, which creates further scaling (size) problems. The dimension differences are illustrated in data furnished by American Electric Power Company for its Cook nuclear unit (1,100 MW) and its Mountaineer coal unit (1,300 MW). The area (square feet) required for turbogenerators, after approximate scaling up of Cook's 1,100 MW to 1,300 MW, is one-third greater for nuclear. Per MW of output, Cook must produce 3.66 times as many cubic feet of steam per hour as Mountaineer.

In sum, the conventional and reactor circuits of nuclear power units have experienced substantial scaling risks. These risks account for a major part of the troubles besetting nuclear power plants of all vintages and for the resulting industry-capacity factors, averaging between 53 percent and 60 percent. The literature provides little indication as to how these problems

Table 2–2
Pressures and Temperatures of Coal and Nuclear Power Units

	Coal				Nuclear				
	Manatee	*Wansley*	*Mansfield*	*Amos*	*Browns Ferry*	*Trojan*	*Beaver V.*	*St. Lucie*	*Indian Point-3*
MW	863	952	914	1,300	1,152	1,216	923	850	1,125
Year Commercial	1976	1976	1976	1973	1974	1976	1976	1976	1976
Type[a]					BWR	PWR	PWR	PRW	PWR
Turbine									
a. PSI	2,400	3,500	3,500	3,500	950	873	735	750	715
b. °F	1,000	1,000	1,000	1,000	575	533	517	513	507
c. RPM	3,600	3,600	3,600	3,600	1,800	1,800	1,800	1,800	1,800
Boilers									
a. Number	1	1	1	1	2	—	—	—	—
b. PSI	2,500	3,625	3,785	3,500	1,005	895	781	750	—
c. °F	1,000	1,000	1,000	1,000	575	533	517	513	507

Source: Adapted from FPC/FERC, *Statistics of Steam Electric Plants*.
[a]BWR = boiling-water reactor.

will be resolved so that capacity factors can rise closer to the original design of 80 percent.

Design-Technological Risks Other than Scale

This study focuses on newly ordered units for service after 1985. Analyses of past casualties such as the study done recently for the Electric Power Research Institute by S.M. Stoller Corporation, may or may not be useful for predicting design-technological risks of future units.[3] Such analyses may not indicate, for example, whether the best state of the art for welding is up to the lifetime requirements of the nuclear containment vessel. They do not indicate to what degree the best state of the art of metallurgy meets the minimum requirements of the plant. Most important, they do not translate the technological information into nuclear economics.

The risks considered here include, but are not limited to, the following for nuclear power plants:

Corrosion in the reactor circuit

Corrosion in the conventional circuit

Welding

Metallurgy in general

Tubing, piping, and walls: thickness and metallurgy

Valves, pumps, and meters: quality and proven adequacy

Controls: automaticity, sensitivity, redundancy, and design

Safety: a combinaton of the preceding factors

The literature on design-technological risks of nuclear power in the United States, France, England, and Germany is identified in the sections that follow.

United States

The United States has a voluminous record of specific outages in the reports required of operators and published by the Nuclear Regulatory Commission (NRC). But the fundamental issues concerning state-of-the-art technology can be extracted only in part, if at all, from the NRC reports, from proceedings of industry meetings, and from publications. In any case, the

United States has devoted little attention to design risks in terms of economics.

Two important design-technological problems have recently received attention in the United States—embrittlement of reactor vessels with age and steam-tube leaks in PWRs—which "may be (the industry's) most serious difficulties yet."[4] The NRC is monitoring eight reactors with brittle pressure vessels and forty reactors with damaged steam-generator tubes.

The embrittlement problem centers on several technological factors, namely,

The degree of thermal shock that a reactor vessel can withstand before cracking

The rate of embrittlement

The effect of cooling water under high pressure on an embrittled vessel

The embrittlement risk is that cooling water might fracture a vessel with tiny cracks. The danger with steam-generator tube leaks (caused by corrosive water) is the release of radioactive steam.

France

France, like the United States, has relatively little literature of the kind found in England and Germany on the fundamental design risks of nuclear power. Based on a political decision to become independent of imported oil and gas and to minimize dependence on coal, the French placed in 1976 a single order for twenty-two replicated nuclear power plants. The twenty-two plants were to be built in a preset sequence to provide the most economical construction cost. Further large orders have been placed since. While there is fairly strong questioning of nuclear power by environmental groups, the French government is continuing with an enlarged program of nuclearization. In 1979, on the initiative of the unions working on twenty-six of the large reactors, important defects were found on the stainless steel coating of the reactor-vessel nozzles and on the steam-generator tube sheets.[5]

Great Britain

The British have had a public soul searching on nuclear-power technologies for some twenty-five years. They have successively committed themselves to the Magnox, the advanced gas-cooled reactor (AGR), the steam-generating heavy-water reactor (SGHWR), the AGR (again, in January 1978), and the pressured-water reactor (PWR). The PWR was authorized conditionally in January 1978, with orders finally awarded in 1979–1980. Hope still exists in some quarters for the high-temperature reactor (HTR) but without any commitment to research or to an operating unit.

Professional unions, the Energy Minister, the Atomic Energy Agency (AEA), the British Central Electricity Generating Board (BCEGB), the South of Scotland Electricity Board (SSEB), the equipment manufacturers, and the constructors have all been in shifting alliances for or against specific technologies. The official reports afford valuable insights into the science and art of such factors as welding, containment vessels, piping, and the safety of light-water reactors (LWR).

Reliability. The "key commercial factor" in choosing a nuclear or fossil-fuel power unit is, according to the British, reliability. In its report of 1974 to Parliament, "Choice of Thermal Reactor Systems," the Nuclear Power Advisory Board stated:

> The Electricity Boards have laid particular stress on the need to base their main requirements on tried and reliable systems. . . . Lack of reliability can rapidly erode any apparent difference in construction costs between reactor systems. . . . As an example, the CEGB estimate that with the AGRs the cost of fuel to replace the electricity production lost through delays will exceed the original estimated cost. (p. 12)

The board then points out the problems of scaling and extrapolating from existing sizes: "If parameters such as pressures and temperatures are increased or new materials used, this will introduce fresh risk which cannot be assessed with confidence until there is operating experience" (p. 12).

Safety and Technology. When the secretary for energy (SFE) decided in 1978 on immediate orders for two twin AGRs, he also promised future orders for PWRs provided that clearance be given on the safety factor. Earlier in December 1976, the Select Committee on Science and Technology of the House of Commons, in its report, *The SGHWR Programme,* wrote:

> In our 1974 report we . . . drew attention to the conflict of expert opinion on the safety of American light water reactors (LWR), and recommended . . . one of the British nuclear technologies which is already proven, since this is likely to satisfy the Nuclear Inspectorate without undue delay. (p. 18)

In England, safety seems to have been of greater concern in reactor choice than competitive costs. This is partly explained by the lack of cost data to sustain a debate. Thus safety has been at the heart of the PWR alter native. In 1976 a two-part study of safety was undertaken, one on the pressure vessel to be done by Dr. Walter Marshall of the U.K. AEA, and a second on other aspects to be done by the Nuclear Installations Inspectorate (NII).

The Marshall report stated that the PWR's pressure vessel could be safe provided it was modified to higher U.K. standards, higher even than in France and Germany. Problems concerned the integrity of the reactor vessel from cracks, metal fatigue, and behavior during transients. This proviso was substantial and found dissenters.

Sir Alan Cottrell, former chief scientific adviser to the government, wrote to the SFE of his concern with the PWR, particularly with regard to senescence. The Marshall report, he observed in *The SGHWR Programme,*

> argues that if very great care is taken, these pressure vessels can be reliable at the start of service, but it offers no certainties for the later years when effects such as metal fatigue may weaken the structure. (p. 35)

He questioned whether "the outstanding technical excellence" of engineering and operating staffs could be obtained. Marshall had been told by ultrasonic operators that defects might not be detectable unless over 1 inch deep. "Yet," Cottrell noted, "this depth is comparable with the critical crack size calculated for some emergency and fault conditions" (p. 35).

Degradation with age also troubled Cottrell; knowledge of both longterm changes in properties of materials and of the critical sizes of defects was insufficient. The AEA shared this concern: "The information on the laws covering crack growth rates is not fully satisfactory from a scientific point of view" (p. 35). The AEA recommended thorough research on corrosion-fatigue crack rates.

The interrelation of technology and human limitations was addressed by Cottrell. He was skeptical of achieving the degree of human perfection needed for the PWR. The SSEB took Cottrell's skepticism as confirmation of their own view that "to have a safe vessel it is essential to fabricate it well nigh perfectly, from well nigh perfect materials, and carry out well nigh perfect inspection throughout its life" (p. 35-36).

The PWR alternative was questioned by the Institution of Professional Civil Servants, whose members are scientists and administrators advising the minister for energy on policy. To the Select Committee, the institution expressed its concern that pressure-vessel failure was "a credible eventuality" (p. 36). It felt that the greater population density of England required more stringent safety features than in the United States. The safety standards of the PWR cooling system did not meet the minimum standards required of the SGHWR. The institution recommended that the NII report on the German Reactor Safety Commission's reasons for requiring additional safety precautions for the PWR at Ludwigshaven, and on the cost.

The need for a secondary shutdown system for a U.K. PWR was raised by the Whetstone Nuclear Power Branch of the Association of Scientific, Technical, and Managerial Staffs. They told the Select Committee that an off-the-shelf PWR was not acceptable to CEGB.

In conversations in the summer of 1977 at the U.K. AEA, it was stated that the United Kingdom does not accept the safety of the Westinghouse PWR. The United Kingdom was particularly concerned with crack propagation in the brittle steel pressure vessels. In light of the United States embrittlement problems, the concerns appeared well founded.

The British government has undertaken an extensive investigation of the safety and economics of coal and nuclear (PWR and AGR) generation. On May 12, 1982, the CEGB presented a twenty-five volume (125 kilogram) study supporting its choice of the PWR. Opponents of this position will present their cases in January 1983. According to the *Financial Times* (London) of May 13, 1982, Cottrell and other "most eminent" skeptics of the PWR are now satisfied with the safety design of "the Achilles heel," the pressure vessel.

Germany

A key question addressed in conversations with the Germans was: To what do they attribute the better performance of German versus U.S. nuclear power plants? Table 2–3 quantifies the answers of Siemens' KWU division, which has manufactured all currently operating nuclear power units in Germany, and of RWE, which is the largest operating utility in the country. Table 2–3 thus represents the seller and buyer viewpoints. Siemens and RWE made their evaluations independently.

KWU assigned 10 percent of the total explanation for its performance differential over the United States to "turnkey" sale. *Turnkey* means that the utility contracts with KWU to design, manufacture, construct, and install the entire plant through testing, for one price and under KWU's sole responsibility.

RWE, however, felt that turnkey has disadvantages that make it less attractive than the alternative; they therefore assigned this factor zero weight. The principal alternative is for RWE to select the optimum combination of component manufacturers and installers from Germany or elsewhere. RWE felt, for example, that Brown-Boveri turbogenerators were preferable to KWU equivalents and perhaps less expensive. In 1977, RWE was working on such an arrangement to serve as general contractor for its next generating units.

Design was a substantial factor in explaining the difference between German and U.S. plant performance. KWU's and RWE's perceptions were

Table 2-3
Weighting of Factors Explaining Higher Capacity Factors and Reliability of German versus U.S. Nuclear Plants

Factors	Weighted by	
	RWE	KWU
1. Turnkey sale	0%	10%
2. Design		
2.1 Reactor design	—	10
2.2 Automation	—	5
2.3 Design details	—	4
2.4 Standardization	—	6
Total	30	25
3. Review of design and procurement	15	20
4. Maintenance representation and service, management, personnel training, preventive maintenance	40	35
5. Regulation and courts	15	10
	100%	100%

fairly close, assigning 25 percent and 30 percent, respectively, of the total weights. KWU further broke down design into four components: reactor: 10 percent; standardization: 6 percent; automation: 5 percent; and other details: 4 percent.

Review of design and procurement was assigned 15 percent and 20 percent, respectively, by RWE and KWU. This included review by the independent TUVs (a nonprofit German engineering-review group) at all stages of the undertaking, which likely involved a more intensive and independent scrutiny than supplied in the United States by the AEC/NRC. The most important single group of factors includes preventive maintenance, factory liaison, management, and personnel training, to which RWE assigned 40 percent and KWU 35 percent.

A fundamental difference in viewpoint (vis-a-vis other countries) is found in German attitudes toward regulations and court intervention. Nuclear power plants in Germany are subject to much more intense regulation than in other countries. There is official scrutiny at all stages: design, siting, manufacture, construction, and operations. RWE estimated that 15 percent of the factors explaining higher capacity factors and reliability is due to "regulations and courts," while KWU estimated 10 percent.

The official inspection process is primarily executed by the Federal Reactor Safety Commission (RSK—Reaktorsicherheitskommission) and the TUVs. The State Minister (which has licensing authority) and the RSK use these nonprofit provincial engineering groups to perform the various inspections and tests. The essential characteristic emphasized for the TUVs is that they are independent.

An important function of the RSK is to sponsor studies of materials used in nuclear power reactors (for example, the study made at the University of Stuttgart on cracks). Material testing has been performed since 1964 and has revealed many serious flaws such as piping walls that are too thin. Dr. Schoch, head of the Baden provincial TUV and of the national TUV (serving as an umbrella for the eleven provincial TUVs), had been proposing for some three and a half years that a special group be established under him to develop a research-testing program for solving problems with materials. This group would contain component producers, steel fabricators, welding specialists, and scientists. Finally, he won his point, and in 1977 the group was created by the Secretary for Technology Research (Forschungsministerium) with a budget of DM 50 million. The goal is to determine the lifetime of material components for nuclear power plants.

It is worth noting a few of the observations made in conversations by Dr. Schoch. He feels (summer of 1977) that the theoretical problems have been solved for nuclear power plants but that the theory has not been translated into equipment and operational performance. The danger, he feels, is that the heads of organizations (for example, reactor safety) are not, as physicists, fully in touch with the real problems in materials, construction, welding, and so on. He observes that the prices of nuclear power plants have been too low from the manufacturer's viewpoint so that suppliers have had to cheapen quality. For example, tubing is too thin; cracks occurred in main feedwater tubing at Phillipsburg before the start of operations. Overall, he feels that safety problems can be solved technologically but that the economic problems of availability and capacity factor are much more difficult to solve.

Conclusion

For some as yet obscure combination of reasons, the U.S. nuclear power industry, the government included, in the late 1950s and early 1960s made a heroic decision to bypass the normal gradualism in the development and commercialization of a new technology—especially an exotic and extraordinarily sensitive technology like nuclear power—and to proceed rapidly to a $100 billion commitment, and ultimately far more. Numerous issues were too little understood, too little researched, and too little acknowledged. The issues included, but were not limited to: fundamental operational questions of the quality and amount of radioactive corrosion; the interrelation of radioactive with conventional corrosion; the practicality of welding art;

the required science and art of metallurgy; the ability of the best available valves, pumps, and meters to meet the demands; the adequacy of the best state of the art of controls; the acceleration of scaling up; human factors (discussed in the next chapter); and the summation of all the technological factors in the end products of capacity factor, reliability, safety, and economics.

Thus, the industry, if it is realistic about the technological and economic success of nuclear power plants, must confront the fundamental pre- and postcommercialization questions that heretofore have been inadequately addressed. In a nutshell, the issue is whether it is possible to design, manufacture, construct, operate, maintain, repair, and decommission nuclear power plants in a safe and economical manner.

In reference to precommerciality risks, the appendix examines evidence for premature commerciality of nuclear power plants. Certain tax and rate-making rules may have encouraged technological risks in order to gain financial benefits.

Notes

1. The European material, unless otherwise specified, was obtained in meetings between one of the authors of this book and industry and government representatives from each country in 1977, 1979, and 1981.

2. FPC/FERC's annual reports provide data by plants, not by units. A plant may consist of one to ten units or more. Therefore, unit data are available only for one unit plant unless the company voluntarily reports by single units (for example, selected for their unusually good heat rates). This limits the number of units available to make the comparisons given here.

3. S.M. Stoller Corp., *Nuclear and Large Fossil Unit Operating Experience,* EPRI NP-1191, September 1979.

4. *Newsweek,* April 19, 1982, pp. 101–102. Steam-generator-tube deterioration is addressed in a March 1980 report by the NRC titled "Evaluation of Steam Generator Tube Rupture Events," NUREG-0651, and the NRC's February 1982 update "Steam Generator Status Report."

5. Bureau of National Affairs, Inc. *Energy Users Report,* October 11, 1979, page 15.

3 Human Factors

An exceptionally high level of quality control is essential at all stages of manufacturing, constructing, and operating nuclear power plants. Quality control depends on labor, management, and design personnel as much as on pure technology. Three limitations operate on the nuclear power industry in terms of human factors:

> The industry must use human beings who, under the best of conditions, are fallible.

> Automation is necessary to mitigate the law of *errare humanum est,* but the fallibility of automation, itself, requires (fallible) human watchdogs.

> The industry is having difficulty attracting and retaining the high quality of personnel required.

It has not yet been determined what degree of automation and human discretion at the control panels of nuclear power plants will minimize error. Siemens in Germany has stated that they prefer higher automation, while the French have stated that they prefer less automation.[1] Because the degree of automation is a function of technological sensitivity and safety, automation is greater in nuclear than in fossil-fuel plants. The frequency of automatic scrams accounts for a good part of the Yoyo Factor (unpredictability of availability) which is discussed in part II.

Human factors, and their interrelationship with automation, played a crucial role in the Three Mile Island accident:

> According to John Kemeny, the former Dartmouth president who headed the official inquiry into the accident, "The real problem was not the equipment but people." Kemeny and his group became convinced that the operators and supervisors at Three Mile Island were negligently unprepared for the events of March 28, 1979. When the water pressure dropped in the steam generator, triggering a chain of other mechanical failures, the technicians, confused and increasingly frightened, misinterpreted gauges, overrode automatic safety systems and nearly caused a catastrophic meltdown.[2]

The three human-factors limitations in the nuclear power industry (identified previously) have interesting parallels in the U.S. military. As much as half of our military's peacetime equipment is not operational due to the lack of skills and intrinsic intelligence to operate, maintain, and repair the equipment. The army has all of its members encoded on a computer by standardized skill modules. When a new weapon is at the approval stage for full-scale production, it is first broken down into corresponding skill modules. The needed skills are then matched with the human supplies available. Not infrequently, the weapon is returned as unacceptable because the supply of human-skill modules does not meet the need. The equipment is then either redesigned within the limits of skill supply or scrapped.

In the early 1960s, the navy was faced with a problem that has already begun to appear in the nuclear power industry and that will become progressively more acute—a low rate of attraction and retention of skilled personnel (in uniform).[3] The *Providence Journal* (August 11, 1980) recently reported a "manpower crisis" in the nuclear power industry because college students are avoiding the industry. Although many jobs in health physics and radiation protection are available, enrollments in these fields have dropped, for example, by two-thirds at the University of Lowell in Massachusetts since 1971.

Nick Catron, an instructor at the TVA-run school for control-room operators, made the following comments about declining morale and job candidates:

> We're losing competent people. . . . They [operators] can make more money on the outside. And if you lose enough experienced people, you're going to have ten people doing jobs that fifteen people are supposed to do. You're going to have longer shifts—and you're going to have trouble. If I were a young man coming into the industry, I would look somewhere else.[4]

The importance of human factors to the technological and hence economic success of the nuclear power industry is discussed in the following pages.

The Best Plants and Human Factors

Germany/Argentina

In Erlangen in 1977, Siemens stated that, of all the plants they had supplied, Atucha in Argentina had the best record. They attributed the superiority chiefly to human factors. When the order was placed, the men to operate and maintain the plant were carefully chosen and sent to Siemens to work in the manufacture and later in the installation and testing of the plant. Because of the high cost, the Argentines have not done this for later plants.[5] The annual capacity factors for Atucha's 5.5-year lifetime through 1979 were successively 40, 68, 69, 41, 78, and 87 percent.[6]

In Germany, a licensing requirement states that a graduate engineer must be present on every shift at a nuclear power plant. Several German sources have stated that the quality and motivation of labor had deteriorated substantially in Germany as well as other Western and Iron Curtain countries.

Spain

In the summer of 1979, Spain's three operating nuclear power plants (with a combined lifetime capacity factor of 69 percent through 1979), had one of the highest, if not the highest, capacity factors of any country. The director of research for the Junta d'Energia Nuclear, Sr. Sevilla, attributed the success to human factors. As with Atucha, the personnel to operate these three plants were preselected carefully and pretrained. Also like Atucha, Sr. Sevilla doubted whether this would be done for subsequent plants because of the expense.

The United States

A perceptive discussion of human factors and nuclear-power operations, held a year before the Three Mile Island-2 accident, is found in an industry meeting on reliability of nuclear power plants held in April 1978.[7] Notes from several of the papers dealing with human factors are given in the following discussion.

Glenn A. Reed, Wisconsin Electric Power Company. In 1952 Reed worked on two prototype nuclear power plants, and since 1956 he has been involved in the initial and continuing staffing of the Yankee Rowe, Haddam Neck, and Point Beach plants. In 1977 he was manager of the Nuclear Power Division of his company. He characterizes these three plants as among the best in the world. Given this background, Reed's analysis of the sources of past and future improvement in the reliability of nuclear power plants is noteworthy.

First, he cites "people performance" as the major reason for the "glaring" differences in reliability among nuclear power plants. People performance involves selection, training, utilization, and motivation.

Second, speaking of technology, he makes the remarkable assertion that improvement of nuclear power has reached a plateau and, except for steam generators, "not much improvement in unit availability in the future should be expected from this quarter."

Third, he characterizes the current (1977) state of human-factors management in the industry as backward. "People training and development aspects were not in good shape in the early 1960s. This aspect strengthened about 1970 and then fell behind again about 1972, and still remains that way to a degree presently."

From the preceding it can be concluded that Reed feels that future improvement in reliability of nuclear power depends mainly on human factors. The published transcript contains much detail that is worth studying.

Richard H. Graves, Haddam Neck plant. In explaining the performance of Haddam Neck, which he characterizes as one of the world's most successful, Grave's first three reasons all involve human factors—initial selection of personnel, slow promotion so that employees know their jobs well, and motivation. His fourth reason is maintenance, an important part of which is highly skilled people and planning. The fifth involves certain reactor-circuit equipment favored by Haddam Neck's management. But of the secondary circuit, he says, "Quality in the design, manufacture, and installation of conventional power plant equipment is definitely lacking. . . . Perhaps an initial investment of a few million dollars in equipment of better quality would have saved millions of dollars in replacement power costs."

Drew C. Smith, Peach Bottom Atomic Power Station. Smith identifies several mechanisms for improving reliability of nuclear power plants—human factors first, and then equipment and skilled maintenance:

> I believe that the goal of increased plant availability can be achieved by a qualified and motivated plant staff supported in fact (not just words) by an enlightened corporate management. The most important items in any plant are people. Without the people, the most sophisticated equipment in the world is just so much junk. The attitude, skills, and training can make or break any operation. The other cornerstone of an operation with high availability is reliable equipment. This should be obtained through a program of preventive maintenance as well as performance and failure review.

Conclusion

Human factors are both an important contributor to nuclear power's performance gap and a key avenue for future improvements. The following questions should be carefully examined.

Demand. What are the requirements, quality and quantity, for personnel in the nuclear power industry?

Supply

What is the supply of such personnel in the national population?

What are the competing demands in other sectors for the same supply?

How well can the nuclear power industry compete for the needed personnel?

Training. What initial and ongoing training should be given to the personnel?

Cost. How much will it cost to attract, train, and retain this supply of manpower?

The design corollary. How do we design and redesign nuclear-power systems to operate at the desired standards, and what are the costs?

Notes

1. The European material, unless otherwise specified, was obtained in meetings between industry and government representatives from each country and one of the authors of this book in 1977, 1979, and 1981.

2. *Life,* May 1982, p. 40. Reprinted with permission.

3. To address the problem, in December 1964 the secretary of the navy formed a task force on navy/marine corps retention of uniformed personnel.

4. *Life,* May 1982, p. 40. Reprinted with permission.

5. Argentina had hired the TUV to advise on Atucha but discontinued halfway through because of the expense.

6. Adapted from basic data in IAEA's *Operating Experience with Nuclear Power Plants in Member States,* annual reports for 1974–1978, and from "Annual Review—1979" in *Nuclear Engineering International,* March 1980. The commerciality dates, nameplate capacity, and nameplate lifetime capacity factors through 1979 for each of the units are as follows: Zorita, August 1969, 161 MW, 72 percent; Garona, May 1971, 464 MW, 61 percent; Vandellos, July 1972, 505 MW, 73 percent.

7. *Reliable Nuclear Power Today,* Proceedings of American Nuclear Society, Piedmont Carolinas and East Carolina Sections, Charlotte, N.C., April 10–13, 1978. Quoted with permission.

4

Learning in the Nuclear Power Industry

The discussion of nuclear power's design/performance gap and the case-study assumptions and results indicate that nuclear power has not performed well to date. This chapter examines one of the most fundamental issues for nuclear-power economics—whether industry performance has improved over time, that is, whether the industry has exhibited learning. The logical assumption that younger units benefit from experience with older units, particularly when similar or identical, is tested empirically.[1]

The performance of similar but different-aged units is analyzed from two perspectives: jumelles and siblings. Jumelles (*twins* in French) are nearly identical but different-aged units that, because they have the following characteristics in common, should offer optimal conditions for learning:

Unit type (for example, boiling-water reactor, pressurized-water reactor)

Unit vendor (for example, Westinghouse, General Electric, Combustion Engineering)

Unit nameplate capacity rating (differing by 10 percent or less in several cases)

Physical site

Operating utility

The sibling analysis examines units that are closely related but not identical. The siblings are the same type, have the same vendor and approximately the same nameplate rating, but do not necessarily share the same site and are not necessarily operated by the same utility.

The purpose of the analysis is to examine the pattern of learning for jumelles and siblings. One would expect positive learning by vendors and operators for both analyses, with perhaps greater learning for jumelles. The methodology for analyzing learning is outlined below, followed by a presentation of the results. (The data problems discussed in the appendix apply to the analysis here as well.)

Methodology

Jumelles Analysis

The jumelles analysis comprised four steps. First, all U.S. units that qualified as twins were organized into twenty-two pairings. For example, Zion-1 and -2 constituted one pair as both are pressurized-water reactors, built by Westinghouse, with nameplate capacity ratings of 1,098 MW, located on the same site, and operated by Commonwealth Edison. For three-unit plants (Browns Ferry and Oconee), the first and second units constituted one pair, the second and third units another, and the first and third another. Nonadjacent units (that is, first and third units) have more opportunity for learning than adjacent units (for example, first and second units) as a longer period of time has elapsed. The twenty-two pairings are described in table 4–1.

Two exceptions to the jumelles characteristics just outlined should be noted. Arkansas-1 and -2 have different vendors (Babcock and Wilcox, and Combustion Engineering, respectively) and Indian Point-2 and -3, while on the same site and initiated by the same utility, have different operating utilities (Consolidated Edison Company and the Power Authority of the State of New York, respectively). These exceptions were not considered significant enough to warrant deleting the two pairings. It should be noted that operation of Brunswick-2 preceded Brunswick-1.

In the second step, learning by jumelles was analyzed for four performance measures:

Elapsed days, that is, the number of days elapsed from a unit's initial criticality to its commerciality

Capacity factor for the first twelve months of a unit's commercial operation

Capacity factor for calendar year 1979 (the most recent calendar year for which data were available)

Lifetime (cumulative) capacity factors through December 1979

The relative performance of younger and older twins was classified into one of three learning categories:

Learning. A significant improvement in performance

No learning. No significant change in performance

Negative learning. A significant deterioration in performance

Table 4-1
Description of Jumelles Units by Pairs

Unit Name	Operating Utility	Nameplate MW	Type Vendor[a]	Commercial Date
1. Arkansas-1	Arkansas Power & Light	903	P/B	12-19-74
Arkansas-2	Arkansas Power & Light	959	P/C	3-26-80
2. Browns Ferry-1	Tennessee Valley Authority	1,152	B/G	8-1-74
Browns Ferry-2	Tennessee Valley Authority	1,152	B/G	3-1-75
3. Browns Ferry-2	Tennessee Valley Authority	1,152	B/G	3-1-75
Browns Ferry-3	Tennessee Valley Authority	1,152	B/G	3-1-77
4. Browns Ferry-1	Tennessee Valley Authority	1,152	B/G	8-1-74
Browns Ferry-3	Tennessee Valley Authority	1,152	B/G	3-1-77
5. Brunswick-2	Carolina Power & Light	867	B/G	11-3-75
Brunswick-1	Carolina Power & Light	867	B/G	3-18-77
6. Calvert Cliffs-1	Baltimore Gas & Electric	918	P/C	5-8-75
Calvert Cliffs-2	Baltimore Gas & Electric	911	P/C	4-1-77
7. Cook-1	Indiana & Michigan Electric	1,152	P/W	8-27-75
Cook-2	Indiana & Michigan Electric	1,133	P/W	7-1-78
8. Dresden-2	Commonwealth Edison	828	B/G	8-11-70
Dresden-3	Commonwealth Edison	828	B/G	10-31-71
9. Hatch-1	Georgia Power Company	850	B/G	12-31-75
Hatch-2	Georgia Power Company	809	B/G	9-5-79
10. Indian Point-2	Consolidated Edison Company	1,013	P/W	8-15-73
Indian Point-3	Power Authority of State of N.Y.	1,013	P/W	8-30-76
11. North Anna-1	Virginia Electric Power Co.	980	P/W	6-6-78
North Anna-2	Virginia Electric Power Co.	947	P/W	12-14-80
12. Oconee-1	Duke Power Company	934	P/B	7-16-73
Oconee-2	Duke Power Company	934	P/B	9-9-74
13. Oconee-2	Duke Power Company	934	P/B	9-9-74
Oconee-3	Duke Power Company	934	P/B	12-16-74
14. Oconee-1	Duke Power Company	934	P/B	7-16-73
Oconee-3	Duke Power Company	934	P/B	12-16-74

Table 4-1 continued

Unit Name	Operating Utility	Nameplate MW	Type Vendor[a]	Commercial Date
15. Peach Bottom-2	Philadelphia Electric Company	1,152	B/G	7-5-74
Peach Bottom-3	Phildelphia Electric Company	1,152	B/G	12-23-74
16. Point Beach-1	Wisconsin Electric Power Co.	524	P/W	12-21-70
Point Beach-2	Wisconsin Electric Power Co.	524	P/W	10-1-72
17. Prairie Island-1	Northern States Power Co.	593	P/W	12-16-73
Prairie Island-2	Northern States Power Co.	593	P/W	12-21-74
18. Quad Cities-1	Commonwealth Edison	828	B/G	8-16-72
Quad Cities-2	Commonwealth Edison	828	B/G	10-24-72
19. Surry-1	Virginia Electric Power Co.	848	P/W	12-22-72
Surry-2	Virginia Electric Power Co.	848	P/W	4-3-73
20. Three Mile-Island-1	Metropolitan Edison Co.	871	P/B	9-2-74
Three Mile Island-2	Metropolitan Edison Co.	961	P/B	12-30-78
21. Turkey Point-3	Florida Power & Light	760	P/W	12-4-72
Turkey Point-4	Florida Power & Light	760	P/W	9-7-73
22. Zion-1	Commonwealth Edison	1,098	P/W	12-31-73
Zion-2	Commonwealth Edison	1,098	P/W	9-19-74

Sources: See table A-4 in appendix.

[a]The first letter indicates unit type (P = pressurized-water reactor, B = boiling-water reactor); the letter after the slash indicates the vendor B = Babcock & Wilcox, C = Combustion Engineering, G = GE, W = Westinghouse).

Table 4–2
Definition of Learning Categories

	Performance Measure	
	Elapsed Days	*Capacity Factor*
Learning	Decrease ≥ 10.0 %	Increase ≥ 5.0 %
No learning	Increase or decrease < 10.0%	Increase or decrease < 5.0 %
Negative learning	Increase ≥ 10.0 %	Decrease ≥ 5.0 %

The quantitative definitions of the preceding categories for each of the four performance measures are given in table 4–2. If the younger twin decreased the time between criticality and commerciality by at least 10 percent, the performance is classified as "learning." A less than 10 percent change in elapsed days is classified as "no learning," and a 10 percent or greater increase as "negative learning."

For capacity factors, learning is defined as a 5-percent (not percentage points) or greater improvement by the younger twin. A less than 5-percent change in either direction is classified as "no learning." If capacity factors decreased by 5 percent or more, the performance is classified as "negative learning."

The third step calculated what percent of the comparisons fell into each learning category. The final step converted the percentages from step 3 into a summary index of the learning pattern. The learning index assigns a weight of + 1 to learning, zero to no learning, and – 1 to negative learning.

The learning index is calculated by multiplying the fraction of comparisons in each learning category by their respective weights and summing the results. For example, if 80 percent of the elapsed days comparisons fell into the learning category, 15 percent into the no-learning category, and 5 percent into the negative-learning category, the learning index calculation would be as follows:

$$(0.80)(1) + (0.15)(0) + (0.05)(-1) = +0.75$$

On this scale, an index near + 1 indicates an overall pattern of learning. An index near zero (roughly between + 0.5 and – 0.5) indicates an overall pattern of no learning. An index near – 1 indicates an overall pattern of negative learning. In the preceding example, an overall pattern of learning would be indicated.

Sibling Analysis

The methodology for the sibling analysis was conceptually identical to the four-step methodology for the jumelles analysis. Step 1 developed five "families" of units, each of which comprised units of the same type, with the same vendor, and with approximately the same nameplating rating. The families are identified below:

General Electric BWRs of 809–883 MW

Westinghouse PWRs of 848–923 MW

Westinghouse PWRs of 980–1,216 MW[2]

Babcock & Wilcox PWRs of 870–963 MW

Combustion Engineering PWRs of 812–959 MW

The units in each family were grouped into "generations" on the basis of commerciality dates. The disaggregation of families into generations is necessarily approximate. For generations with more than one unit (as was generally the case), an average was taken for each performance measure. For example, the elapsed days for the second generation of Babcock & Wilcox PWRs constituted an average of the elapsed days for the five units in that generation. Table 4–3 describes the units in each generation for the five families.

In step 2, intergenerational performance was compared from two points of view: comparisons were made between adjacent generations (that is, one generation compared with the immediately preceding generation) as well as between nonadjacent generations in each family. For example, the analysis of General Electric BWRs compared the average elapsed days for the following generations: second versus first, third versus second, third versus first, fourth versus third, fourth versus second, and fourth versus first. Again, nonadjacent units have more opportunity to learn. As with the jumelles analysis, the relative performance of younger and older generations was classified into one of three learning categories—learning, no learning, or negative learning—according to the definitions in table 4–2.

The third and fourth steps were identical for the sibling and jumelles analyses. The percent of comparisons that fell into each learning category was calculated and these percentages were converted into the learning index that summarizes the overall pattern of learning.

The jumelles and sibling comparisons analyzed fifty-one U.S. nuclear units (half of which were used in both comparisons). The fifty-one units constituted 81 percent of the sixty-three operating U.S. nuclear units (in 1980) with nameplate ratings of 500 MW or more.

Table 4-3
Description of Sibling Units by Family

Generation	Unit Name	Operating Utility	Nameplate (MW)	Commercial Date
Family 1 GE BWRs (809–883 MW)				
First	Dresden-2	Commonwealth Edison	828	8-11-70
Second	Dresden-3	Commonwealth Edison	828	10-31-71
	Quad Cities-1	Commonwealth Edison	828	8-16-72
	Quad Cities-2	Commonwealth Edison	828	10-24-72
Third	Cooper Station	Nebraska Public Power Dist.	836	7-1-74
	Fitzpatrick	Power Authority of State of N.Y.	883	7-28-75
	Hatch-1	Georgia Power Company	850	12-31-75
	Brunswick-2	Carolina Power & Light	867	11-3-75
Fourth	Brunswick-1	Carolina Power & Light	867	3-18-77
	Hatch-2	Georgia Power Company	809	9-5-79
Family 2 Westinghouse PWRs (848–923 MW = Medium)				
First	Surry-1	Virginia Electric Power Co.	848	12-22-72
	Surry-2	Virginia Electric Power Co.	848	4-3-73
Second	Beaver Valley-1	Duquesne Light	923	9-30-76
	Farley-1	Alabama Power Company	888	12-1-77
Family 3 (Westinghouse PWRs (980–1,216 MW = Large)				
First	Indian Point-2	Consolidated Edison Co.	1,013	8-15-73
	Zion-1	Commonwealth Edison	1,098	12-31-73
	Zion-2	Commonwealth Edison	1,098	9-19-74
Second	Cook-1	Indiana & Michigan Electric Co.	1,152	8-27-75
	Trojan	Portland General Electric Co.	1,216	12-24-75
Third	Indian Point-3	Power Authority of State of N.Y.	1,013	8-30-76
	Salem	Public Service Electric & Gas	1,170	6-30-77
Fourth	North Anna-1	Virginia Electric Power Company	980	6-6-78
	Cook-2	Indiana & Michigan Electric Co.	1,133	7-1-78
	North Anna-2	Virginia Electric Power Company	947[a]	12-14-80

Table 4-3 continued

Generation	Unit Name	Operating Utility	Nameplate (MW)	Commercial Date
Family 4 Babcock & Wilcox PWRs (870–963 MW)				
First	Oconee-1	Duke Power Company	934	7-16-73
Second	Three Mile Island-1	Metropolitan Edison Company	871	9-2-74
	Oconee-2	Duke Power Company	934	9-9-74
	Oconee-3	Duke Power Company	934	12-16-74
	Arkansas-1	Arkansas Power & Light	903	12-19-74
	Rancho Seco	Sacramento Municipal Utility Dist.	963	4-17-75
Third	Crystal River-3	Florida Power Company	890	3-13-77
	Davis Besse	Toledo Edison Company	962	12-31-77
	Three Mile Island-2	Metropolitan Edison Company	961	12-30-78
Family 5 Combustion Engineering PWRs (812–959 MW)				
First	Palisades	Consumers Power Company	812	12-31-71
	Maine Yankee	New England Electric Power Co.	864	12-28-72
Second	Calvert Cliffs-1	Baltimore Gas & Electric	918	5-8-75
	Millstone-2	Northeast Utilities Company	910	12-26-75
Third	St. Lucie-1	Florida Power & Light	850	12-21-76
	Calvert Cliffs-2	Baltimore Gas & Electric	911	4-1-77
Fourth	Arkansas-2	Arkansas Power & Light	959	3-26-80

Sources: See table A–4 in appendix.

[a] North Anna-2 is classified as a large PWR because it has the same design-electrical rating and thermal-power rating as North Anna-1, which has a 980 MW nameplate rating.

Results

The percentage of the jumelles comparisons that falls into each learning category is summarized in table 4-4.[3] (Data for each pair and an explanation of missing comparisons are given in tables 4-8 and 4-9 at the end of the chapter.) The four performance measures together, constituting eighty-one comparisons, do not indicate an overall pattern of learning. Slightly less than half of all jumelles comparisons fall into the learning category and slightly more than half fall into the no- and negative-learning categories combined. More comparisons show negative (32 percent) than no learning (22 percent). Only the first-twelve-months capacity factor exhibits a predominant learning pattern—approximately 60 percent of the comparisons show learning, compared with 40 percent in the no/negative-learning categories combined. The elapsed-days and cumulative-capacity-factor comparisons each fall roughly equally into the learning and no/negative-learning catagories. Approximately twice as many 1979 capacity-factor comparisons show no/negative learning as learning.

The distribution of the sibling comparisons by learning category is summarized in table 4-5.[4] (Data for each generation and an explanation of missing comparisons are given in tables 4-10 and 4-11 at the end of the chapter.) The eighty-two adjacent and nonadjacent comparisons combined indicate no overall pattern of learning. Overall, more comparisons show no/negative learning than learning, and more comparisons show negative than no learning. Only the first-twelve-months-capacity-factor comparisons fall predominantly into the learning category. For each performance measure, comparisons in the negative-learning category always outnumber or are approximately equal to comparisons in the no-learning category.

Table 4-4
Jumelles Comparisons in Each Learning Category

		Capacity Factors			
	Elapsed Days	First 12 Months	1979	Lifetime	Total
Number of Comparisons	21	22	19	19	81
Learning (%)	42.9	59.1	31.6	47.4	45.7
No learning (%)	9.5	22.7	15.8	42.1	22.2
Negative learning (%)	47.6	18.2	52.6	10.5	32.1
No and negative learning (%)	57.1	40.9	68.4	52.6	54.3

Table 4–5
Sibling Comparisons in Each Learning Category

| | Elapsed Days | Capacity Factors | | | |
		First 12 Months	1979	Lifetime	Total
Adjacent Generations					
Number of comparisons	12	12	11	11	46
Learning (%)	16.7	58.3	18.2	27.3	30.4
No learning (%)	33.3	8.3	27.3	36.4	26.1
Negative learning (%)	50.0	33.3	54.5	36.4	43.5
No and negative learning (%)	83.3	41.7	81.8	72.7	69.6
Nonadjacent Generations					
Number of comparisons	10	10	8	8	36
Learning (%)	10.0	70.0	12.5	37.5	33.3
No learning (%)	20.0	10.0	25.0	25.0	19.4
Negative learning (%)	70.0	20.0	62.5	37.5	47.2
No and negative learning (%)	90.0	30.0	87.5	62.5	66.7
All Comparisons					
Number of comparisons	22	22	19	19	82
Learning (%)	13.6	63.6	15.8	31.6	31.7
No learning (%)	27.3	9.1	26.3	31.6	23.2
Negative learning (%)	59.1	27.3	57.9	36.8	45.1
No and negative learning (%)	86.4	36.4	84.2	68.4	68.3

The hypothesis that nonadjacent generations are more likely to exhibit learning is not borne out by the overall or individual performance-measure results. The adjacent and nonadjacent comparisons reveal remarkably similar overall learning patterns. The distribution of total comparisons across learning categories is nearly identical for adjacent and nonadjacent comparisons. Both have a majority of comparisons in the learning category for the first-twelve-months capacity factor only. Nonadjacent generations performed worse than adjacent generations for elapsed days and the 1979 capacity factor and better than adjacent generations for the first-twelve-months and cumulative-capacity factors.

Table 4–6 compares and combines the jumelles and sibling results. The two analyses together, comprising 163 comparisons, do not indicate a pattern of learning. Sixty-one percent of the total comparisons show no or negative learning. Again, the first-twelve-months-capacity factor is the only measure for which comparisons in the learning category outnumber comparisons in the no/negative-learning categories. The cumulative-capacity factor is the only measure that has more comparisons in the no- than in the negative-learning category.

A comparison of the jumelles and sibling results reveals very similar learning patterns for the two analyses. Neither demonstrates an overall pattern of learning although the jumelles performed slightly better than the siblings. Both exhibit a predominant learning pattern for the first-twelve-months-capacity factor only. While neither analysis exhibits learning for the other three performance measures, the jumelles demonstrate more learning than siblings. Negative learning occurs more frequently than no learning for two of the performance measures in the jumelles analysis. Negative learning outnumbers no learning for all measures except the cumulative-capacity factor in the sibling analysis.

The learning-index results are presented in table 4–7. The summary learning index for the jumelles and sibling analyses combined falls squarely on no learning (+0.001). The summary index for the jumelles analysis (+0.136) is better than for the sibling analysis (−0.134) but both summary indexes indicate no learning. The summary indexes for the individual performance measures all indicate no learning with the first-twelve-months-capacity factor having the highest index.

Looking at the individual performance measures for the jumelles and siblings separately, all of the indexes indicate no learning. The elapsed-days and 1979 capacity-factor measures for siblings fall close to negative learning (−0.455 and −0.421, respectively), and the first-twelve-months-capacity factor for jumelles falls close to learning (0.409). The jumelles, as expected, have consistently better indexes than the siblings.

Several caveats should be kept in mind. First, the lifetime capacity factor incorporates the first-twelve-months- and 1979-capacity factors, causing some, but generally not significant, overlap across the measures. Each of the three capacity factors presents different, albeit related, perspectives on performance. Second, overlap also exists between the jumelles and sibling analyses. The two analyses, though using some of the same units, provide different perspectives on learning. Third, the numbers that combine performance measures and/or siblings and jumelles assign equal weight to the components being combined. The analysis has been presented in disaggregated as well as aggregated form in light of these caveats.

Table 4-6
Summary of Comparisons in Each Learning Category

	Elapsed Days			Capacity Factors First 12 Months			1979			Lifetime			Total		
	Jumelles	Sibling	Total	Jumelles	Sibling	Total	Jumelles	Sibling	Total	Jumelles	Sibling	Total	Jumelles	Sibling	Total
Number of comparisons	21	22	43	22	22	44	19	19	38	19	19	38	81	82	163
Learning (%)	42.9	13.6	27.9	59.1	63.6	61.4	31.6	15.8	23.7	47.4	31.6	39.5	45.7	31.7	38.7
No learning (%)	9.5	27.3	18.6	22.7	9.1	15.9	15.8	26.3	21.1	42.1	31.6	36.8	22.2	23.2	22.7
Negative learning (%)	47.6	59.1	53.5	18.2	27.3	22.7	52.6	57.9	55.3	10.5	36.8	23.7	32.1	45.1	38.6
No and negative learning (%)	57.1	86.4	72.1	40.9	36.4	38.6	68.4	84.2	76.3	52.6	68.4	60.5	54.3	68.3	61.3

Table 4–7
Learning-Index Results

	Elapsed Days	Capacity Factors			
		First 12 Months	1979	Lifetime	Total
Jumelles analysis	− 0.047	+ 0.409	− 0.21	+ 0.369	+ 0.136
Sibling analysis	− 0.455	+ 0.363	− 0.421	− 0.052	− 0.134
Total	− 0.256	+ 0.387	− 0.290	+ 0.158	+ 0.001

The above caveats not withstanding, the results of the jumelles and sibling analyses clearly indicate that, even under optimal learning conditions, the gap between design and performance has not been narrowing and may, in some respects, be widening. Jumelles appear to learn more than siblings, and the best learning period is the first twelve months of commercial operation.

Table 4-8
Data for Jumelles Analysis

Pairing (First and Second Unit)	Elapsed Days		Capacity Factors (Percent)					
			First 12 Months		1979		Lifetime	
	First Unit	Second Unit	First Unit	Second Unit	First Unit	Second Unit	First Unit	Second Unit
1. Arkansas-1&2	135	477	61.6	61.4	a	a	a	a
2. Browns Ferry-1&2	349	224	38.8	16.3	74.2	73.8	43.5	43.6
3. Browns Ferry-2&3	224	205	16.3	70.7	73.8	54.4	43.6	59.0
4. Browns Ferry-1&3	349	205	38.8	70.7	74.2	54.4	43.5	59.0
5. Brunswick-2&1	228	161	38.9	46.6	48.1	41.8	44.4	50.9
6. Calvert Cliffs-1&2	213	122	70.7	79.2	52.2	68.8	63.2	69.5
7. Cook-1&2	221	113	62.7	64.8	56.1	60.0	56.7	60.8
8. Dresden-2&3	216	273	29.5	64.8	68.1	47.9	52.7	51.7
9. Hatch-1&2	475	428	55.1	60.6	a	a	a	a
10. Indian Point-2&3	85	146	17.5	79.4	54.1	54.0	41.4	59.5
11. North Anna-1&2	62	185	68.4	66.1	a	a	a	a
12. Oconee-1&2	88	302	54.3	56.0	61.2	72.9	55.4	57.9
13. Oconee-2&3	302	102	56.0	60.7	72.9	39.8	57.9	59.4
14. Oconee-1&3	88	102	54.3	60.7	61.2	39.8	55.4	59.4
15. Peach Bottom-2&3	292	138	65.7	52.7	85.0	60.5	60.6	57.9
16. Point Beach-1&2	49	124	71.0	45.9	66.6	80.8	69.4	73.6
17. Prairie Island-1&2	15	4	28.0	59.2	56.0	80.7	60.0	68.2
18. Quad Cities-1&2	303	181	67.4	69.6	65.9	54.9	56.7	56.2
19. Surry-1&2	174	27	47.1	64.5	30.4	8.3	52.1	49.5
20. Three Mile Island-1&2	89	277	77.0	15.7	11.1	15.6	58.6	15.6
21. Turkey Point-3&4	45	88	49.7	61.3	43.2	57.7	58.4	57.9
22. Zion-1&2	195	269	36.3	42.7	57.6	49.4	52.8	55.4

aSecond unit too new to make comparison.

Table 4-9
Incidence of Learning in Jumelles Comparisons

| Twins Compared | Elapsed Days | | Capacity Factors | | | | | | |
| | | | First 12 Months | | 1979 | | Lifetime | |
	Percent Change	Learning Category	Percent Change	Learning Category	Percent Change	Learning Category	Percent Change	Learning Category
1. Arkansas-2 vs. 1	+253	Negative	0	No	a	a	a	a
2. Browns Ferry-2 vs. 1	−36	Learning	−58	Negative	−0.5	No	+0.2	No
3. Browns Ferry-3 vs. 2	−9	No	+334	Learning	−26	Negative	+35	Learning
4. Browns Ferry-3 vs. 1	−41	Learning	+82	Learning	−27	Negative	+36	Learning
5. Brunswick-1 vs. 2	−29	Learning	+20	Learning	−13	Negative	+15	Learning
6. Calvert Cliffs-2 vs. 1	−43	Learning	+12	Learning	+32	Learning	+10	Learning
7. Cooks-2 vs. 1	−49	Learning	+3.3	No	+7	Learning	+7	Learning
8. Dresden-3 vs. 2	+26	Negative	+120	Learning	−30	Negative	−1.9	No
9. Hatch-2 vs. 1	−10	No	+10	Learning	a	a	a	a
10. Indian Point-3 vs. 2	+72	Negative	+354	Learning	−0.2	No	+44	Learning
11. North Anna-2 vs. 1	+198	Negative	−3.4	No	a	a	a	a
12. Oconee-2 vs. 1	+243	Negative	+3.1	No	+19	Learning	+4.5	No
13. Oconee-3 vs. 2	−66	Learning	+8	Learning	−45	Negative	+2.6	No
14. Oconee-3 vs. 1	+16	Negative	+12	Learning	−35	Negative	+7	Learning
15. Peach Bottom-3 vs. 2	−53	Learning	−20	Negative	−29	Negative	−4.5	No

Table 4-9 continued

| | Capacity Factors | | | | | | | |
| | Elapsed Days | | First 12 Months | | 1979 | | Lifetime | |
Twins Compared	Percent Change	Learning Category	Percent Change	Learning Category	Percent Change	Learning Category	Percent Change	Learning Category
16. Point Beach-2 vs. 1	+153	Negative	−35	Negative	+21	Learning	+6	Learning
17. Prairie Island-2 vs. 1	b	b	+111	Learning	+44	Learning	+14	Learning
18. Quad Cities-2 vs. 1	−40	Learning	+3.3	No	−17	Negative	−0.9	No
19. Surry-2 vs. 1	−84	Learning	+37	Learning	−73	Negative	−5.0	Negative
20. Three Mile Island-2 vs. 1	+211	Negative	−80	Negative	+41c	Noc	−73	Negative
21. Turkey Point-4 vs. 3	+96	Negative	+23	Learning	+34	Learning	−0.9	No
22. Zion-2 vs. 1	+38	Negative	+18	Learning	−14	Negative	+4.9	No

Note: This table indicates, in percentage terms, how the younger twin performed relative to the older twin. Percentages of 6.0 or greater have been rounded to the nearest whole number. Percentages less than 6.0 have been rounded to the nearest decimal place. The plus and minus signs indicate whether the percentage change was positive or negative.

aData were available for the younger unit to make the comparison.

bThe elapsed days for Prairie Island-1 and -2 (fifteen and four days, respectively) were considered to be extreme anomalies and were not therefore included in the comparisons.

cWhile in 1979 Three Mile Island-2 performed significantly better in percentage terms than Three Mile Island-1, their relative performance was classified as "no learning" due to the very low absolute values of their 1979 capacity factors (15.6 percent versus 11.1 percent).

Table 4–10
Data for Sibling Analysis

	Generation	Period	Number of Units	Average Elapsed Days	Average Capacity Factors		
					First 12 Months	Calendar 1979	Lifetime
Family 1							
GE	First	1970	1	216	29.5	68.1	52.7
BWRs	Second	1971–72	3	252	67.3	56.3	54.9
	Third	1974–75	4	272	48.8	49.9	51.0
	Fourth	1977–79	2	294	53.6	41.8	50.9
Family 2							
Westinghouse	First	1972–73	2	100	55.8	19.3	50.8
Medium PWRs	Second	1976–77	2	128	51.3	22.2	37.6
Family 3							
Westinghouse	First	1973–74	3	183	32.2	53.7	49.8
Large PWRs	Second	1975	2	115	40.8	52.8	47.3
	Third	1976–77	2	173	57.9	37.0	46.6
	Fourth	1978–80	3	120	66.4	54.4	59.5
Family 4							
Babcock & Wilcox	First	1973	1	88	54.3	61.2	55.4
PWRs	Second	1974–75	5	168	56.0	46.7	56.7
	Third	1977–78	3	159	37.4	33.7	32.4
Family 5							
Combustion Engineering	First	1971–72	2	144	34.8	54.1	48.4
PWRs	Second	1975	2	142	63.4	54.1	60.5
	Third	1976–77	2	182	74.9	67.2	68.7
	Fourth	1980	1	477	61.4	a	a

[a]Unit too new to make comparison.

Table 4-11
Incidence of Learning in Sibling Comparisons

Generations Compared	Elapsed Days		Capacity Factor First 12 Months	
	Percent Change	Learning Category	Percent Change	Learning Category
Family 1				
GE BWRs				
2nd vs. 1st	+17	Negative	+128	Learning
3rd vs. 2nd	+8	No	−27	Negative
3rd vs. 1st	+26	Negative	+65	Learning
4th vs. 3rd	+8	No	+10	Learning
4th vs. 2nd	+17	Negative	−20	Negative
4th vs. 1st	+36	Negative	+82	Learning
Family 2				
Westinghouse Medium PWRs				
2nd vs. 1st	+28	Negative	−8	Negative
Family 3				
Westinghouse Large PWRs				
2nd vs. 1st	−37	Learning	+27	Learning
3rd vs. 2nd	+51	Negative	+42	Learning
3rd vs. 1st	−4.8	No	+80	Learning
4th vs. 3rd	−31	Learning	+15	Learning
4th vs. 2nd	+4.3	No	+63	Learning
4th vs. 1st	−34	Learning	+106	Learning
Family 4				
Babcock & Wilcox PWRs				
2nd vs. 1st	+91	Negative	+3.1	No
3rd vs. 2nd	−6	No	−33	Negative
3rd vs. 1st	+80	Negative	−31	Negative
Family 5				
Combustion Engineering PWRs				
2nd vs. 1st	−1.4	No	+82	Learning
3rd vs. 2nd	+29	Negative	+18	Learning
3rd vs. 1st	+27	Negative	+115	Learning
4th vs. 3rd	+161	Negative	−18	Negative
4th vs. 2nd	+237	Negative	−3.2	No
4th vs. 1st	+232	Negative	+76	Learning

Capacity Factors

Generations Compared	1979		Lifetime	
	Percent Change	Learning Category	Percent Change	Learning Category
Family 1				
GE BWRs				
2nd vs. 1st	−17	Negative	+4.1	No
3rd vs. 2nd	−11	Negative	−7	Negative
3rd vs. 1st	−27	Negative	−3.3	No
4th vs. 3rd	−16	Negative	0	No
4th vs. 2nd	−26	Negative	−7	Negative
4th vs. 1st	−39	Negative	−3.4	No
Family 2				
Westinghouse Medium PWRs				
2nd vs. 1st	+15[a]	No[a]	−26	Negative
Family 3				
Westinghouse Large PWRs				
2nd vs. 1st	−1.8	No	−5.1	Negative
3rd vs. 2nd	−30	Negative	−1.5	No
3rd vs. 1st	−31	Negative	−7	Negative
4th vs. 3rd	+47	Learning	+28	Learning
4th vs. 2nd	+3.1	No	+26	Learning
4th vs. 1st	+1.3	No	+19	Learning
Family 4				
Babcock & Wilcox PWRs				
2nd vs. 1st	−24	Negative	+2.3	No
3rd vs. 2nd	−28	Negative	−43	Negative
3rd vs. 1st	−45	Negative	−41	Negative
Family 5				
Combustion Engineering PWRs				
2nd vs. 1st	0	No	+25	Learning
3rd vs. 2nd	+24	Learning	+14	Learning
3rd vs. 1st	+24	Learning	+42	Learning
4th vs. 3rd	b	b	b	b
4th vs. 2nd	b	b	b	b
4th vs. 1st	b	b	b	b

Note: This table indicates, in percentage terms, how younger generations performed relative to older generations. Percentages of 6.0 or greater have been rounded to the nearest whole number. Percentages less than 6.0 have been rounded to the nearest decimal place. The plus and minus signs indicate whether the percentage change was positive or negative.

[a] While in 1979 the second generation of Westinghouse Medium PWRs performed significantly better in percentage terms than the first generation, their relative performance was classified as "no learning" due to the very low absolute values of their 1979 capacity factors (22.2 percent versus 19.2 percent).

[b] The fourth generation was too young to have the data necessary for the comparisons.

Notes

1. An analysis of learning for fossil units, while desirable, cannot be made as individual-unit data are not available.

2. See footnote in table 4–3 concerning megawatt rating for North Anna-2.

3. An exception to the learning-category definitions in table 4–3 should be noted. The Three Mile Island comparison was classified as no learning for the 1979 capacity factor even though on a percentage basis it fell into the learning category. The absolute value of the capacity factors (15.6 percent versus 11.1 percent) was so low as to warrant an exception to the definitions.

4. The exception to the learning-category definitions explained in note 3 also applied to the family 2 comparison for the 1979 capacity factor. The absolute values of the capacity factors were 22.2 percent versus 19.3 percent in this instance.

**Part II
Methodology and
Assumptions for Case Studies**

5

Introduction to Part II

Overview

The methodology and assumptions used in the economic analysis of nuclear and coal power are addressed in part II. This chapter explains the case-study approach and summarizes the specific adjustments used in the four case studies. Chapters 6 through 9 identify and discuss analytical issues fundamental to an adequate economic study of nuclear and coal power. The omission and/or inadequate treatment of such issues has been a major shortcoming of published economic studies to date. To the extent feasible, the issues are translated into quantitative assumptions to be used in the four case studies. The probable impacts of unquantified assumptions are also discussed.

Analytical Approach

The Four Case Studies

The analysis begins with four case studies of nuclear and coal economics:

Atomic Energy Commission (AEC), *The Nuclear Industry 1974,* WASH 1174–74

Energy Research and Development Administration (ERDA), *Comparing New Technologies for the Electric Utilities, Draft Final Report,* December 9, 1976, ERDA 76–141

Nuclear Regulatory Commission (NRC), *Draft Environmental Statement, New England Power Units 1 and 2,* May 1979, NUREG-0529

Exxon Research and Engineering Company (Exxon), *An Outline for a Discussion on the Economics of Nuclear Power with Dr. Richard Hellman of the University of Rhode Island,* November 18, 1977, and subsequent conversations in 1979

The first authoritative statement of nuclear and coal generating costs was issued in February 1974 by the AEC in its annual review, *The Nuclear Industry 1973*. For the first time, assumptions were clearly stated and sensitivity analysis was performed. Because the price of oil had just quadrupled on January 1, 1974, to $10/barrel (from $2.25/barrel FOB Ras Tanura, Saudi Arabia), the AEC recognized that price relationships for coal, oil, and nuclear fuels were in a state of upheaval. The dust had settled somewhat by the following year so that the 1974 report was chosen for the analysis here. The latter is current as of November 15, 1974, though issued about May 1975. (After issue of the report, the AEC was dissolved into NRC and ERDA.)

The ERDA study of December 1976 compares the feasibilities of various existing and potential technologies for the generation of electricity to the years 2000 and 2020. The primary focus is "the economics of power production" and "the strategy of commercialization." It was intended as "a step in achieving a dialogue between ERDA, which sponsors the development of new technology, and the private sector, which would implement the newly developed technologies." While the division of inputs is not clearly defined, the authors were ERDA and two firms, TRW and the METREK Division of MITRE. The *Draft Report* cited previously, managed by a joint ERDA/TRW/MITRE task force, also remained the final version. Information on the ERDA study was provided from three other sources as listed here:

TRW, Electric Utilities Study, *An Assessment of New Technologies from a Utility Viewpoint-Final Report,* October 30, 1976, prepared for ERDA

MITRE/METREK Division, *Analysis of Benefits Associated with the Introduction of Advanced Generating Technologies: Description of Methodologies and Summary of Results,* March 1977, L. Goudarzi, Project Leader, prepared for ERDA

L. Goudarzi, further data on the study transmitted by letter to Professor Richard Hellman, December 5, 1980; and by telephone to Professor Richard Hellman, January 12, 1981, and October 1981.

The NRC study analyzed the competitive economics of a proposed nuclear generating plant in Charlestown, Rhode Island. The study was performed by the NRC staff as part of its environmental-impact statement. A critique of the study was made in 1979 by one of the authors of this book at the request of Rhode Island's governor.

The Exxon study was made by its Research and Engineering Company, which is staff for the entire corporation, as a guide to top management on its future posture toward nuclear power. The Nuclear Division of Exxon did not participate in the study. While the study is private, it was made available to one of the authors of this book.

All four studies analyze the economics of nuclear and coal units newly ordered for initial commercial operation in the 1980s. AEC and Exxon developed initial commerciality costs, and ERDA and NRC analyzed lifetime costs. All case studies examined large (that is, 1,000 MW or greater) nuclear and coal units, except for NRC, which examined medium-size (767 MW) coal units. The coal units use high-sulfur Appalachian or Midwest coal with flue-gas desulfurization (FGD). Exxon also examined costs for low-sulfur Appalachian coal without FGD. All of the coal units satisfied federal pollution standards in effect at the time of the estimates.

Exxon's low-sulfur-coal scenario is the best option over a ten-year horizon for new coal-burning plants. Low-sulfur coal is still preferable when the costs of FGD are added to comply with the Environmental Protection Agency's New Source Performance Standards (NSPS). Low-sulfur (less than 1 percent) "recoverable" reserves of bituminous coal in the Appalachian region comprise 14.5 billion tons. The 1.1 billion tons in the Midwest give a total of 16 billion tons, which is half the "demonstrated" reserves. At the projected use of about 540 million tons of coal by electric utilities in 1990, these reserves constitute a twenty-nine-year supply of low-sulfur coal.[1] All the reserves would not be used by the utilities since much of it is metallurgical coal.

The time horizon for evaluating the supply of low-sulfur coals is bounded by the probability that within five to ten years, raw burning of coal with FGD will become obsolete. Improved cleaning and preparation and processes such as fluidized bed combustion (FBC) and coal gasification will likely make the use of high-sulfur coal economic with little or no FGD. Thus it is economically feasible to use the low-sulfur Appalachian and Midwest coals now, with the understanding that new processes allowing large use of high-sulfur coals will be commercialized within five to ten years. As far as economically practical, new coal-burning plants should be built with provision in the designs for conversion to FBC or other processes at a later date.

The selection of the four case studies was based on two factors. First they represent four of the best existing studies of comparative nuclear and coal economics. Only their sponsors (or similar organizations) have the resources and authority to obtain the cost estimates from the most valid sources, the architect/engineers and manufacturers of power plants. The three government studies were performed by competent and knowledgeable teams with official access to outside sources. The Exxon study represents the vast resources, competence, and knowledge of the world's largest energy company. Second, each sponsor is neutral, if not sympathetic toward nuclear power.

While the four case studies are the best available, they all have considerable shortcomings. Simply deciphering methodologies and assumptions was often a difficult task. The ERDA study was the most poorly and confusingly presented. It is worth describing several problems with the study that required the assistance of the project leader, Mr. Goudarzi, to unravel.

A useful economic comparison of nuclear and coal power, in the end, expresses the major cost components (that is, capital, fuel, O&M, and their sum) in mills/kwh. ERDA, however, provides only O&M and total costs in mills/kwh. Capital and fuel costs are given in measures that are not additive. Further, it is unclear whether ERDA's extraordinarily complex assumptions and methodologies are used consistently in the main report and in the two companion documents (cited previously).

ERDA's capacity-factor treatment illustrates the report's inconsistencies and ambiguities. No specific capacity factors are given in the body of the draft report, fundamental as they are to comparative costs. The draft report and the MITRE documents imply that the same capacity factor was assumed for nuclear and coal. The draft report states:

> The baseload technologies were assumed to have similar plant factor[2] histories over lifetimes—typically starting at 0.5, rising to 0.7 after 2 or 3 years, and later steadily decreasing until the plant is retired. In effect, this assumption involves treating all technologies as having comparable reliability regardless of their inherent complexity. (p. 102)

And the MITRE report states:

> The capacity factor profile for each of the technologies was chosen in a manner that would allow all technologies to compete on a roughly equivalent economic basis in terms of the amount of utilization of capacity (ie, similar capacity factors) but one that also recognized technological differences (eg, conventional technologies would likely experience fewer break-in problems than advanced technologies; hence the former would have a higher initial capacity factor). (p. 51)

In the report, nuclear and coal were both classified as "conventional."

The MITRE report goes on to assert that nuclear will achieve a higher lifetime capacity factor than coal. No average lifetime capacity factors are given but first-, third-, and thirtieth-year capacity factors are: 60, 72, and 50 percent, respectively, for nuclear and 60, 70, and 30 percent, respectively, for coal. The coal-capacity factors are identical to those found in the appendix of the draft report, which specifies a lifetime capacity factor of 60 percent. In fact, as was learned from Mr. Goudarzi, the ERDA study assumed 65 percent for nuclear and 58 percent for coal. No lifetime capacity factor for nuclear can be found in either the draft report or the MITRE report.

Thus the capacity-factor information is incomplete, inconsistent, and incorrect.

Inconsistent numbers are also given for fuel and total busbar costs. The following fuel costs are specified for coal in cents per million Btu (no usable fuel costs are provided for nuclear):

Source	1975	2000
Draft report	89–119¢	111–149¢
TRW report	53–94	70–135
MITRE report	77–104	100–135

The following total lifetime busbar costs are provided in mills/kwh:

Source	Nuclear	Coal
Draft report, p. 195	17.8 mills	21.8 mills
Draft report, p. 20	18.5	21.6
Draft report, p. A–2	—	26.9
Draft report, p. 153	—	25.0
TRW Report	34.4	35.0
Goudarzi	17.2	21.3

Adjustments to Case Studies

Once the methodologies and assumptions of the four case studies were understood, the first task was to make them comparable by developing lifetime costs for AEC and Exxon, and by converting NRC's three medium-size coal units to two large units. The second task was to correct the most important shortcoming of the four case studies: the omission of key economic factors and the unrealistic treatment of factors that were included. Realistic assumptions are developed, quantified, and incorporated into the four case studies for the following economic factors:

Capital-cost assumptions
    Technological sufflation
    Construction time
    Economic life of units
    Capacity factor

Fuel-cost assumptions
    Btu/pound of coal

Heat rate

Operating-and-maintenance (O&M) cost assumptions.[3]

Economic factors that are qualitatively assessed but, due to methodological and substantive difficulties, could not be quantified and incorporated into the case studies, include the following:

The Yoyo Factor

Major repairs

Replacement power

Senescence

Fuel prices

Use of low-sulfur coal (quantified for Exxon only)

Waste disposal

Decommissioning

The adjustments required for the four case studies indicate that, while they represent the best studies to date, their economics were inadequate as originally presented. This inadequacy is striking, given the magnitude of the U.S. commitment to nuclear power of well over $100 billion. The NRC and Exxon studies are the most reliable of the four because they are more recent and sounder methodologically than the AEC and ERDA studies.

Because of the greatly differing methodologies and assumptions used in the original case studies, direct comparison of costs across studies is not feasible. The analysis focuses therefore on the ratios of coal to nuclear costs, which can be validly compared. The absolute values of the cost projections should in any case be viewed as approximations, given the general uncertainties inherent in long-range prediction and the specific uncertainties with nuclear-power predictions. The cost ratios are approximately valid for units going commercial in the 1990s as well as the 1980s.

The analytical issues and assumptions are assessed, quantitatively or qualitatively, in chapters 6 through 9. The general methodology for incorporating each quantified factor into the case studies is explained. (The calculations are detailed for each case study in part III.) The following section summarizes the quantified adjustments to the case studies.

Summary of Adjustments. The "base case" comprises the original studies adjusted for the realistic assumptions developed here:

1. Technological sufflation: a 30-percent increase in the projected capital costs for nuclear in all four case studies
2. Construction time: for nuclear, 12 years in NRC and Exxon, 10 years in AEC, and 10.5 years in ERDA (the differences reflect what could have been best predicted at the time the study was made, as opposed to hindsight); for coal, 4.5 years in all four case studies
3. Unit life: for nuclear, 25 years in all four case studies; for coal, 30 years in all four case studies
4. Capacity factor: for nuclear, 55 percent in all four case studies; for coal, 70 percent in all four case studies
5. Btu/pound of coal: 12,000 Btu/lb in all four case studies
6. Heat rate: 9,200 Btu/kwh for high-sulfur coal in all four case studies; 9,050 Btu/kwh for low-sulfur coal in Exxon
7. Operating and maintenance costs: for nuclear, the O&M actually achieved by large nuclear units in 1978, adjusted for the time period of each case study

Unless otherwise indicated, tables and numbers refer to the base case.

In addition to the base-case assumptions just outlined, four sensitivity analyses were performed (for lifetime costs only) assuming higher and lower capacity factors for nuclear and coal. Table 5-1 summarizes the cases analyzed.

Before proceeding, the terms used extensively throughout the case study analyses are defined briefly below:

Table 5-1
Summary of Cases Analyzed

	Capacity Factor Assumed	
	Nuclear	Coal
Base case	55%	70%
N = 60% case	60	70
N = 50% case	50	70
C = 75% case	55	75
C = 65% case	55	65

Note: The base case was developed for lifetime costs in all four case studies, for initial commerciality costs in AEC, and for first-year costs in Exxon. The remaining cases (that is, the sensitivity analysis) were developed for only lifetime costs in all four case studies.

Case study refers to the AEC, ERDA, NRC, and Exxon studies.

Scenario refers to the use of high-sulfur coal with FGD or low-sulfur coal without FGD.

Original case study, assumptions, and so on refers to the case studies and so on before adjustments are applied.

Case refers to a set of capacity factors for coal and nuclear incorporated into the original case studies (see table 5–1 for definitions). The cases analyzed include:

Base case

$N = 60$ percent case

$N = 50$ percent case

$C = 75$ percent case

$C = 65$ percent case

Notes

1. Derived from President's Commission on Coal, *Coal Data Book,* 1980.

2. *Plant factor* and *capacity factor* are synonymous.

3. Operating and maintenance costs (O&M) refer, throughout the book, to nonfuel O&M costs.

6 Capital Costs

Estimated capital costs per unit of output for a future power plant are based on two important assumptions—total capital costs (including equipment and financing costs) and the number of kwh over which total costs are averaged. Of the four capital-cost adjustments made to the case studies, the first two correct for unrealistic total cost assumptions, and the second two correct for unrealistic output (that is, kwh) assumptions. The technological-sufflation adjustment corrects the overly optimistic equipment estimates for nuclear and for the resulting underestimation of financing costs. A second adjustment corrects nuclear and coal financing costs for unrealistic construction times. The third and fourth adjustments—unit life and capacity factor—correct unrealistic assumptions about the number of kwh over which the equipment and financing costs are averaged. Each of the adjustments is discussed in separate sections that follow.

Before proceeding, an aspect of capital cost that has received little detailed attention in print should be mentioned. The dramatically increased ratio of fuel to total cost for fossil-fuel generation encourages greater capital intensity for greater fuel economy. In late 1973 coal represented about 30 percent of total generation costs for a newly ordered plant. Today it is approximately 70 percent of the total costs. Capital/fuel tradeoffs apply to new nuclear and fossil-fuel plants as well as to existing plants with renovation possibilities.

The economics of capital-fuel tradeoffs are illustrated by a Tennessee Valley Authority (TVA) program initiated in 1978. TVA devoted $197 million from 1979 through 1981 for boiler rehabilitation and turbine repair on nine coal units built in the 1950s. As a result of the $197 million investment in plant improvements, TVA expects to save $75 million annually, $50 million of which is attributable to improved unit availability and $25 million of which is due to heat-rate improvements (the latter representing a savings of 700,000 tons of coal).

Technological Sufflation

Total capital costs must take into account two unrelated components of cost inflation:

Pure price inflation, which should generally apply equally to nuclear and coal costs

Technological sufflation,[1] which results from technological uncertainty at the design stage of power plants

The galloping rise in nuclear power plant costs relative to the coal-plant costs has been due preponderantly to technological sufflation. Nuclear power's low capacity factors relative to design and the related concerns for safety have in part resulted from regulatory mandated "fixes" and constraints on future construction and operation. How then can realistic future capital costs be developed?

Equipment costs (independent of inflation) can be estimated in several ways. Past costs can be extrapolated using regression analysis. For nuclear power plants, this method is limited by an inadequate data base and by inconsistent and widely scattered numbers for past units. Overall it affords a view of the past, more than a basis for estimating equipment costs of a future nuclear plant.

The second estimation method, and most widely used, is to obtain estimates for a given technology and time period from power-plant suppliers. Such estimates for coal have generally been on target and can therefore be used with reasonable confidence. Supplier estimates for nuclear, however, have invariably resulted in unit performance below design and costs well above the initial contract estimates. That is, technological sufflation has not been taken into account.

The third method adjusts nuclear design estimates obtained from plant suppliers to a desired and realistically achievable level of performance. This is the approach used here with the four case studies.

The four case studies illustrate well the assertion that capital-cost estimates ($/kw) by suppliers tend to be on target for coal and to be underestimated for nuclear. Tables 6–1 through 6–6 compare the original $/kw estimates in the case studies to estimates for units with similar commerciality dates found in security prospectuses of 1981–1982.

The pattern of comparisons provides a useful perspective on the general accuracy of the case-study estimates. Precise numerical comparisons of the capital costs are limited for several reasons. First, the commerciality dates are similar but not always identical across units in a grouping. Second, the megawatt ratings are based on varying capacity definitions, for example, nameplate, design electrical rating. Third, the estimates assume different construction times and rates for interest during construction (IDC) and price escalation during construction (EDC). Fourth, the prospectuses do not always clearly indicate whether the first fuel loading is included in the nuclear-cost numbers. Finally, while the nuclear units compared are all large, medium-sized coal units had to be used to widen the base for comparison.

Several low-sulfur-coal units in the prospectuses did not include FGD. In order to make them comparable with the four case studies with FGD, 20 percent was added to capital costs for FGD. The 20 percent increment is based on current estimates obtained in discussions (in July 1982) with the staffs of the Electric Power Research Institute (EPRI), the Tennessee Valley Authority (TVA), the American Electric Power Company (AEP), and the Environmental Protection Agency (EPA); the estimates ranged from 20 to 30 percent, with low-sulfur coal at the lower end of the range. The 20-percent increment for low-sulfur coal is supported by Arkansas Power and Light's cost estimates (security prospectus of January 27, 1982) for Independence-1 and -2 (low-sulfur-coal units, commercial in 1983 and 1985) with and without FGD. The identical units with FGD and the same coal were estimated to be 16 percent more expensive than without FGD.

As tables 6–1 through 6–3 indicate, the capital-cost estimates ($/kw) for coal in the four case studies are well within the ballpark of estimates for actual coal units. Capital costs for the three units to which the AEC is compared (table 6–1) average 8 percent higher than the AEC's estimate, with a range of 20 percent cheaper to 22 percent more expensive. The three units average 9 percent cheaper than ERDA's estimate (table 6–1), with a range of 32 percent cheaper to 3 percent more expensive.

The NRC is compared to six coal units (table 6–2), which average 8 percent more expensive than the NRC's capital-cost estimate, with a range of 27 percent cheaper to 46 percent more expensive. Table 6–3 compares Exxon's capital-cost estimates for high- and low-sulfur coal to six units. The six units with FGD average 9 percent cheaper than Exxon's high-sulfur coal estimate, with a range of 21 percent to 1 percent cheaper. The six units without FGD averaged 3 percent higher than Exxon's low-sulfur coal estimate without FGD, with a range of 8 percent cheaper to 12 percent more expensive.

The capital-cost comparisons for the nuclear units (tables 6–4 through 6–6) contrast sharply with coal results. The case-study estimates are consistently and considerably below the average capital costs for the comparison units. The capital costs for nine nuclear units to which the AEC is compared (table 6–4) average 118 percent higher than the AEC's estimate with a range of 18 percent to 333 percent more expensive. The nine units average 79 percent higher than ERDA's estimate, ranging from approximately the same as ERDA's to 256 percent above it.

The NRC is compared to three nuclear units (table 6–5) which average 189 percent more expensive than the NRC, with a range of 86 to 277 percent. Capital costs for the ten units to which Exxon is compared (table 6–6) average 72 percent higher than Exxon's nuclear estimate, ranging from nearly identical to Exxon to 182 percent higher.

The summary data in table 6–7 illustrate the important points of the capital-cost comparisons. First of all, the case-study estimates for nuclear are clearly underestimated, and for coal are clearly close to target. The

Table 6–1
AEC and ERDA Case Studies: Coal Capital-Cost Comparisons (Including FGD)

Unit	Coal Type[a]	Sulfur Content[b]	Year Commercial	Size (MW)	$/kw	Capital Cost (with FGD)[c]	
						AEC = 100	ERDA = 100
AEC	Bit.	High	1982	1,000	570	100	—
ERDA	Bit.	High	1982	1,000	673	—	100
Sandow	Lig.	Low	1981	545	455	80	68
Gibson-5	Bit.	Mid	1982	650	695	122	103
Independence-1[d]	Sub-B.	Low	1984	816	691	121	103
Average (3 units)						108	91

Sources: Adapted from security prospectuses for (and follow-up conversations with) the following companies: (1) Texas Power and Light, September 10, 1981, for Sandow data; (2) Public Service of Indiana, April 29, 1982, for Gibson data; and (3) Arkansas Power and Light, January 27, 1982, for Independence data.

Note: The capital costs are to be used for general comparison only as explained in the text.

[a]Bit. = bituminous; Lig. = lignite; Sub-B. = sub-bituminous.

[b]High = 3 percent and above; Low = 1 percent and below; Mid = between 1 and 3 percent.

[c]The comparison units all had FGD in the original estimates.

[d]The Independence cost represents an average of the plant's two units.

Table 6-2
NRC Case Study: Coal Capital-Cost Comparisons (Including FGD)

Unit	Coal Type[a]	Sulfur Content[b]	Year Commercial	Size (MW)	Capital Cost (with FGD)[c]	
					$/kw	NRC = 100
NRC[d]	Bit.	High	1988	1,150	902	100
Twin Oak-1 & 2	Lig.	Low	1986&1988	750	793	88
Forest Grove	Lig.	Low	1989	750	1,074	119
Miller-3	Bit.	Low	1989	660	1,219	135
Martin Lake	Lig.	Low	1990	750	654	73
Miller-4	Bit.	Low	1991	660	1,318	146
Average (6 units)						108

Sources: Adapted from security prospectuses for (and follow-up conversations with) the following companies: (1) Texas Power and Light, September 10, 1981, for Twin Oak, Forest Grove, and Martin Lake data; and (2) Alabama Power Company, April 23, 1981, for Miller data.

Note: The capital costs are to be used for general comparison only as explained in the text.

[a]Bit. = bituminous; Lig. = lignite.

[b]High = 3 percent and above; Low = 1 percent and below.

[c]The Miller units were the only comparison units without FGD in the original estimates. Twenty percent was added to the Miller costs for FGD as explained in the text.

[d]As explained in chapter 13, the NRC originally assumed three 767-MW coal units that were converted here to two 1,150-MW units.

Table 6-3
Exxon Case Study Coal Capital-Cost Comparisons

Unit	Coal Type[a]	Sulfur Content[b]	Year Commercial	Size (MW)	Capital Costs with FGD[c] $/kw	Exxon Hi-S = 100	without FGD[c] $/kw	Exxon Lo-S = 100
Exxon	Bit.	High	1985	1,000	875	100	—	—
Exxon	Bit.	Low	1985	1,000	—	—	645	100
Belle River- 1 & 2	Sub-B.	Low	1984-85	675	756	86	630	98
Rockport-1 & 2	Bit.	Low	1984&1986	1,300	866	99	722	112
Miller-2	Bit.	Low	1985	660	829	95	691	107
Independence-2[d]	Sub-B.	Low	1984	816	691	79	594	92
Average (6 units)					91		103	

Sources: Adapted from Security prospectuses for (and follow-up conversations with) the following companies: (1) Detroit Edison Company, April 1, 1982, for Belle River data; (2) American Electric Power, April 7, 1981, for Rockport data; (3) Alabama Power Company, April 23, 1981, for Miller data; and (4) Middle South Utilities, October 29, 1981, for Independence data.

Note: The capital costs are to be used for general comparison only as explained in the text.

[a]Bit. = bituminous; Sub-B. = sub-bituminous.

[b]High = 3 percent and above; Low = 1 percent and below.

[c]The original estimates for the comparison units did not include FGD. Twenty percent is added to the costs for the Exxon high-sulfur with FGD comparison as explained in the text.

[d]The Independence cost represents an average of the plant's two units.

Table 6-4
AEC and ERDA Case Studies: Nuclear Capital-Cost Comparisons

Unit	Year Commercial	Size (MW)	$/kw	Capital Cost AEC = 100	Capital Cost ERDA = 100
AEC	1982	1,000	680	100	—
ERDA	1982	1,000	828	—	100
Farley-2	1981	860	950	140	115
Salem-2	1981	1,115	800	118	97
Grand Gulf-1	1982	1,250	1,892	278	229
Comanche Peak-1 & 2	1982&1984	1,150	972	143	117
Shoreham	1983	849	2,945	433	356
Waterford-3	1983	1,104	1,426	210	172
Zimmer-1	1983	793	1,576	232	190
Fermi-2	1983	1,100	1,818	267	220
Average (9 units)				218	179

Sources: Adapted from security prospectuses for (and follow-up conversations with) the following companies: (1) Alabama Power Company, April 23, 1981, for Farley data; (2) Philadelphia Electric Company, April 2, 1981, for Salem data; (3) Middle South Utilities, October 29, 1981, for Grand Gulf and Waterford data; (4) Texas Power and Light, September 10, 1981, for Comanche Peak data; (5) Long Island Lighting Company, December 15, 1981, for Shoreham data; (6) American Electric Power, April 7, 1981, for Zimmer data; and (7) Detroit Edison Company, April 1, 1982, for Fermi data.

Note: The capital costs are to be used for general comparison only as explained in the text.

Table 6–5
NRC Case Study: Nuclear Capital-Cost Comparisons

			Capital Cost	
Unit	Year Commercial	Size (MW)	$/kw	NRC = 100
NRC	1988	1,150	1,201	100
Wash PPSS-3	1986 (Dec.)	1,240	3,655	304
Nine Mile Pt.-2	1987	1,083	4,524	377
Pilgrim-2	1987	1,150	2,233	186
Average (3 units)				289

Sources: Adapted from security prospectuses for (and follow-up conversations with) the following companies: (1) Washington Public Power Supply System, September 4, 1981, for WPPSS data; (2) Long Island Lighting Company, December 15, 1981, for Nine Mile Point data; and (3) Public Service Company of New Hampshire, February 1981, for Pilgrim data.

Note: The capital costs are to be used for general comparison only as explained in the text.

Table 6–6
Exxon Case Study: Nuclear Capital-Cost Comparisons

			Capital Cost	
Unit	Year Commercial	Size (MW)	$/kw	Exxon = 100
Exxon	1985	1,000	1,210	100
Seabrook	1984	1,150	1,831	151
Susquehanna-1 & 2	1984	1,050	1,667	138
Wash PPSS-2	1984	1,100	2,924	242
Catawba-1 & 2	1984–85	1,145	1,196	99
Millstone-3	1986	1,150	2,996	248
Wash PPSS-1	1986	1,250	3,414	282
Limerick-1 & 2	1985&1987	1,055	1,953	161
Average (10 units)				172

Sources: Adapted from security prospectuses for (and follow-up conversations with) the following companies: (1) Public Service Company of New Hampshire, February 1981, for Seabrook and Millstone data; (2) Pennsylvania Power and Light, December 9, 1981, for Susquehanna data; (3) Washington Public Power Supply System, September 4, 1981, for WPPSS data; (4) North Carolina Municipal Power Agency # 1, September 17, 1981, for Catawba data; and (5) Philadelphia Electric Company, April 2, 1981, for Limerick data.

Note: The capital costs are to be used for general comparison only as explained in the text.

Table 6–7
Summary of Capital-Cost Comparisons

	Number of Comparison Units	*Case Study = 100*	*Range*	
			Low	*High*
Nuclear				
AEC	⎱ 9	218	118	433
ERDA	⎰	179	97	356
NRC	3	289	186	377
Exxon	10	172	99	282
Average[a] (22 units)		199		
Coal				
AEC	⎱ 3	108	80	122
ERDA	⎰	91	68	103
NRC	6	108	73	146
Exxon Hi-S	⎱ 6	91	79	99
Exxon Lo-S	⎰	103	92	112
Average[a] (15 units)		100		

aThe four case studies are compared with twenty-two nuclear units and fifteen coal units. For weighting purposes, nine nuclear and nine coal units were used twice.

twenty-two comparison units for nuclear average, overall, 99 percent higher than the case studies' units. The fifteen comparison units for coal average (coincidentally) exactly the same costs as the case studies' units.

The second important point of the analysis is that the range of the comparison results differs markedly for nuclear and coal. None of the nuclear comparison units has costs significantly below a case study's costs. Every case study, in contrast, has at least one coal comparison unit with significantly lower costs. The comparison units for nuclear range from 3 percent cheaper than a case-study estimate to 333 percent more expensive; for coal, the comparison units range from 32 percent cheaper to 46 percent more expensive.

The technological-sufflation adjustment is based on the AEC case study which, in its sensitivity analysis, examined a 30-percent increase in capital costs. Given the results of the capital-cost comparisons, no technological-sufflation adjustment is applied to coal. The AEC's 30-percent increase was accepted as a conservative and convenient adjustment for nuclear. It does not include the costs of closing the design/performance gap (that is, improving capacity factors above 55 percent) or of major repairs (see chapter 8).[2]

Construction Time

The construction times in nearly all case studies were too high for coal and too low for nuclear units, even at the time each of the original projections was made. As table 6–8 indicates, the eleven largest fossil-fuel units awarded in 1971 (550 to 1,300 MW) were scheduled to be completed in three to six years, with most falling within four to five years. While none of the units has FGD, the construction time would not generally change with FGD. Installation of FGD requires two to three years, which can be accomplished within the four to five years required for the total plant. (The preceding estimates are based on discussions in July 1982 with the staffs of EPRI and EPA.) Of the eleven units, seven were completed as scheduled, and four took an extra year. However, the coal construction times assumed by AEC, ERDA, and Exxon were 6.5, 6, and 6 years, respectively; NRC assumed approximately 4.5 years, which coincides with reality.

Nuclear-power construction delays, on the other hand, have been substantial. The FPC, concerned with this problem, undertook a survey that found that 27,000 MW of nuclear power were in delay in 1973–1974. General Peters, appointed by the AEC to study solutions for the problem, reported to both the AEC and the nuclear power industry that the entire frame of reference would have to be changed radically, that nuclear power plants could not be built like fossil-fuel plants. Since nuclear power is capital intensive, interest during construction (IDC) and price escalation during construction (EDC) are substantial items.

The adjusted construction time for coal is 4.5 years. Adjusted nuclear construction times were determined for each case study according to what was known at the time the study was made (10, 10.5, 12, and 12 years for AEC, ERDA, NRC, and Exxon, respectively). These construction times for nuclear are conservative—a recent *Life* magazine article (May 1982, p. 40) states that "it can take as long as 14 years to build and start up a plant." The construction-time adjustment is applied to the IDC and EDC components of capital costs.

Economic Life of Units

The technological sufflation and construction-time adjustments correct unrealistic total capital-cost assumptions for coal and nuclear. The next step is to correct the unrealistic output assumptions for average costs beginning with the number of years for which a unit can be expected to operate.

The economic life of nuclear power plants is highly problematic because of the pioneer technologies in the primary radioactive circuit and in the secondary conventional circuit.[3] The radioactive circuit, in addition to

Table 6-8
Scheduled versus Actual Construction Times for Fossil-Fuel Units Announced or Awarded in 1971

Unit	MW	Fuel[a]	Completion Year		Years Scheduled	Overrun in Years
			Announced	Actual		
Gavin-1	1,300	C	1974	1974	3	0
Gavin-2	1,300	C	1975	1975	4	0
Mansfield-1	880	C	1975	1976	4	1
Mansfield-2	880	C	1976	1977	5	1
Manatee-2	825	O	1975	1976	4	1
DeCordova-1	775	G	1975	1975	4	0
Martins Creek-3	800	O/C	1975	1975	4	0
Martins Creek-4	800	O/C	1977	1977	6	0
Monticello-1	550	C	1974	1974	3	0
Monticello-2	550	C	1975	1975	4	0
Canal 2	560	O	1975	1976	4	1

Source: U.S. AEC, *The Nuclear Industry 1971*, p. 83 for dates of announcement and scheduled operation. DOE/FPC, *Steam-Electric Plant Construction Cost and Annual Production Expenses* for 1974–1977 for actual commerciality dates.

[a]C = coal, O = oil, G = gas.

being new and complex, is without historical commercial analog. Also, units have been scaled up in size so fast that there has been little time to accumulate and embody experience with smaller units.

The secondary circuit, though using generically known fossil-fuel technologies, was complicated by two elements of newness. First, operating at approximately half the temperatures, and a third to a fourth of the pressures of fossil-fuel plants, light-water reactors (LWRs) require, for the same output, much larger components such as pumps, valves, piping, and turbogenerators. These sizes have involved extrapolations beyond the experienced limits of scaling for fossil-fuel units of the same size. Second, the scaling problem is further complicated by leaks of radioactivity from the primary into the secondary circuits, with resulting corrosion and other degradations of materials. These problems help explain why the secondary circuits have had more casualties than the primary circuits.

In the United States, the economic life for nuclear power units is usually stated as thirty years. Indications that the economic life is less than thirty years can be found in Europe[4] and in the United States. In Europe the following economic lives are used:

Electricite de France: twenty to twenty-one years.

British Central Electricity Generating Board: twenty years.

RWE, the largest German utility: twenty years based on "internal calculations."

Dr. Schoch, manager of the generating station at Mannheim and head of the national and Baden TUVs, has estimated in conversations that the nuclear-plant life is under twenty years and that a twenty-year life would involve substantial additional expense.

In the United States, significant deviation from the thirty-year assumption is found in the 1975 study done for the New England Power Company (NEPCO) by Arthur D. Little's (ADL) subsidiary, Stoller. The report itself does not state the assumed life of coal and nuclear, but upon request, ADL told NEPCO that a thirty-year life was assumed for a coal plant but twenty-eight years for a nuclear plant. The twenty-eight-year life for nuclear is important not so much for the particular number but because ADL/Stoller felt that nuclear power unit lives would be below the conventional thirty years.[5] AEP estimates the useful life of nuclear and coal units as twenty-five and thirty to thirty-three years, respectively (in its security prospectus of April 7, 1981).

The economic life of a fossil-fuel unit is generally assumed to be thirty years. This number has been amply validated with experience. Fossil-fuel units also have a proven usefulness beyond thirty years as standby and peaking capacity.

Thus, if the economic life for fossil fuel is assumed to be thirty years, the light-water-reactor generating plant, existing or ordered today, seems unlikely to equal it. Given the problems of radioactivity and scaling, an estimate between the twenty years of the French-British-German assumptions and the twenty-eight years of ADL/Stoller is more reasonable than thirty. However, any estimate is precisely that, given the lack of experience on which to base a prediction for nuclear power units. Of the first two U.S. units, Indian Point-1 is permanently shut down. Dresden-1 was shut down on October 31, 1978, with a scheduled return on line in July 1981 but extended to mid-1986 (DOE *Update,* January–March 1982).

The unit size now being ordered is above 1,000 MW, but the oldest such unit did not become commercial until August 1973 (Indian Point-3, at 1,013 MW), providing as of 1982 only nine years of experience. If this size is assumed to be the direct descendant of the 828-MW Dresden-2, which went commercial in August 1970, we have twelve years of experience.

All four case studies assumed an economic life of thirty years for both nuclear and coal units. The coal assumption is not changed, and the nuclear life is adjusted to twenty-five years. Under the shorter life, the same amount of depreciation is recovered but at a greater annual rate. The return on investment (ROI) is an annual constant and taxes tend to be a constant. The chief impact of the shorter life is that capital costs are recovered for fewer kwh so that average costs rise. If it is assumed that the last five years of the thirty-year life average the same capacity factor as the first twenty-five, then the simplest approximation of the increase in mills/kwh due to the shorter life is 30/25ths of the capital component.[6]

Annual Capacity Factor

Economic comparisons of nuclear and coal generation must ultimately focus on average costs, that is, total costs divided by the number of kwh in the period. Unit lifetime, discussed previously, is one determinant of the kwh. Annual capacity factor, however, is the most important single measure of economic comparison between nuclear and coal power.

Because of the confused state of capacity-factor measures in the literature, this section begins with a clarification of key definitions and concepts. An important point concerning nuclear and fossil-fuel capacity-factor comparisons—the *hydro analog*—is then introduced. Senescence, or the decline of annual capacity factors with plant age, is addressed next, followed by the specific capacity-factor adjustment used in the four case studies.

Definitions and Concepts

Capacity factor is defined as follows:

$$\frac{\text{Actual kwh in period}}{\text{(MW capacity)(hours in period)}}$$

All three components of capacity factor have multiple interpretations. Unfortunately, because much public data and many analyses do not clearly state which definitions are being used, the analyst must use extreme caution in interpreting capacity-factor data. Each component is addressed separately in the following discussion.

Kwh. For calculating generation costs, the end-product, that is, net electricity sent out of the plant, is clearly the relevant quantity. In calculating capacity factors, the FERC/FPC has always used net kwh. AEC before 1973 used gross kwh, but has since used net kwh. The International Atomic Energy Agency (IAEA) has used gross kwh in the past but currently uses net kwh. Net kwh should generally be the numerator in calculating capacity factor for economic comparisons.

MW Capacity. The following measures of capacity are found in the industry.

a. Nameplate (NP). NP is the gross MW specified on the brass plate attached to the generator unit by the manufacturer. Since NP is the final output of the entire generating system, there is a logical presumption that the components throughout the system are sized, generally, to the nameplate rating of the generator. These components include the steam turbine, steam generators, piping, valves, and pumps. The nameplate rating generally is permanent, finite, objective, known, warranted by the manufacturer and, above all, logical. None of the alternative capacity measures to be discussed meets these specifications to the same degree. Nameplate is the only capacity-factor measure used by FERC/FPC. In *Statistics of Steam-Electric Plants,* for example, line 1 for each generating plant is nameplate MW, line 2 is net kwh, and line 3 is "plant factor" (that is, capacity factor) derived from lines 1 and 2.

b. Net Maximum Dependable Capacity (NMDC). NMDC is the capacity concept preferred by NRC and the Edison Electric Institute. It is defined as the maximum dependable capacity of a unit, winter or summer, whichever is less. NMDC depends on conditions of cooling water and air temperatures, which are factors in generating efficiency. While this is a useful concept for planning the contribution of a generating unit to power supply, it is not valid as the basic measure for economic comparison of coal and nuclear power. NMDC may be determined by the limitation during a relatively few hours at the worst season of the year. It ignores average capability over the year, availability for a peak that may occur at another time of the year, and the fact that the entire system is designed around the nameplate rating of the generator. In addition to these functional defects, NMDC will fluctuate, from year to year and season to season, with the amount and temperatures of cooling water and with air temperatures. NMDC does not meet the tests of permanence, manufacturer's warranty, and validity for economic comparisons.

c. *Design Electrical Rating, Net (DER).* DER is used by NRC in addition to NMDC. It is the output specified by the ordering utility for plant design. DER and NP for every unit appear in the NRC *Monthly Status Report.* The sum of DER capacities for the sixty-seven units operating on January 1, 1980, was 94.6 percent of their nameplate ratings, which is approximately the average ratio of net to gross kwh. While DER is useful before a unit is completed, thereafter it is obsolete and should not be used for computing capacity factors.

d. *Authorized Power.* This designation is used by IAEA in its annual reports, *Operating Experience with Nuclear Power Plants in Member States.* Beginning with the first report in 1970, variations for the same unit appear from year to year and within the same year, creating considerable confusion and inaccuracies. The erratic IAEA data argue strongly for using nameplate rating exclusively in economic comparisons.

e. *Other Definitions.* There is a broad variety of other capacity measures used in various countries. The British, for example, use "derated" capacity, which is the experienced average capacity. It can vary widely from year to year and within the same year. Derating is useful, for example, in planning power supply and perhaps in labor negotiations.[7] Since derated capacity is the only measure published, its variability is a strong argument for using nameplate rating as the basic measure of economic performance.

Hours in Period. The period definition for annual capacity factors is uniform throughout the data sources. Lifetime capacity factors, however, differ across sources. NRC lifetime capacity factors as of 1982, for example, may begin from first electricity, not from day of commerciality. These dates can differ by months or, as in the case of Ft. St. Vrain, by years.

Other sources state that they begin lifetime capacity factors with commerciality.

The Hydro Analog

A basic technological and economic characteristic of nuclear power is that it has a comparatively low incremental cost per kwh. In Exxon's study, for example, comparing 1,000-MW coal and nuclear units located in New England in the first year of commerciality, the ratio of fuel to total cost (including capital) per kwh is 19 percent for nuclear, 39 percent for Appalachian coal with flue-gas desulfurization (FGD), 47 percent for low-sulfur Appalachian coal without FGD, and 75 percent for low-sulfur fuel oil at $15/barrel (85 percent at $30). Therefore, nuclear should be run as baseload in preference to coal and oil. It is analogous to run-of-stream (ROS) hydroelectric power whose incremental cost is zero. ROS hydro would be used before all other sources of fuel generation, except in unusual circumstances. If the available nuclear power exceeds a utility's demand, it would be advantageous to sell excess nuclear power to adjoining areas to substitute for fossil-fuel generation.

As a general rule, therefore, nuclear power plants should be expected to generate electricity whenever available. The point is clearly illustrated by comparing *service factors* (SF) and *operating availability factors* (OAF) for nuclear. SF is the ratio of hours a generator is actually on-line to period hours. OAF is the ratio of hours a generator is available (whether or not it is on-line) to period hours. Taking the three Browns Ferry units of the TVA, the Fitzpatrick and Indian Point-2 units of the Power Authority of New York, and the Trojan-2 unit in the Pacific NW, representing the three U.S. regions with much hydro (particularly ROS), in every case the SFs are identical with the OAFs in 1979. The corollary is that virtually all nonavailability is forced due to required shutdowns, although the timing of the shutdowns may be somewhat flexible. For all U.S. nuclear units, average SFs, and OAFs were virtually identical for 1978, 1979, 1980, and 1981 (from NRCs *Status Report*). Coal, in contrast, is "load following" and would be expected to have SFs lower than OAFs.

Senescence

Senescence is the decline in annual capacity factor with age of plant. It has had little explication in the literature except for an ERDA memorandum of 1975, and several variations.

Most explicit predictions of capacity factor over time assume a low point during the first year, and a rise to a third-year peak that lasts for the balance of the assumed thirty-year unit life. Alternatively, a falling capacity factor may be assumed after some age. While there is a fairly good historical record by which to gauge senescence of fossil-fuel units, nuclear units are too young and the technology is too immature for definite predictions.

The design-technological risks discussed in part I are likely to differentiate nuclear from fossil-fuel senescence. The scaling differences between nuclear and fossil-fuel units create more uncertainty as to nuclear's unit life. Perhaps more important, radioactivity is highly corrosive; the exact extent over a unit's life is not known, but it has been greater than anticipated. Leakage of radioactivity from the primary into the secondary circuits has become a major problem.

The various sources of information on senescence are discussed separately below.

The 1975 ERDA Memorandum. This unpublished memorandum of February 1975 gives the following sequence of senescence for a nuclear power plant.[8]

Years	Capacity Factor (%)	
1 & 2	65	
3–15	75 high	(The authors of this memo-
	70 low	randum would now use 65,
		see below).
16–30	minus 2% per year to a minimum of 40%	

With the 70-percent plateau, the thirty-year capacity factor is 67.0 percent. Using a 65-percent plateau, the thirty-year capacity factor drops to 63.0 percent. The significance of this memo is not in the specific capacity factors but in the introduction of the senescence factor.

In conversations in the summer of 1980, Snyder, one of the two authors, said that their senescence sequence has become "a classic," and that she was not aware of any disagreement with the pattern (although the sequence is not always used). Bown, the other author, stresses the absence of senescence in the literature, the empirical basis for their sequence, and the lack of history on which to base a better prediction. To his recollection, only one person (at the Pacific Northwest Labs in Richland) has questioned the sequence, suggesting only that the fall-off be less abrupt.

The RWE Comment, 1977. The head of the nuclear division of Germany's largest electric utility, RWE at Essen, observed in conversations that the Bown-Snyder senescence was a correct principle, based on their experience and studies. However, he would start the decline in the eighteenth rather than the sixteenth year.

NRC, 1978 and 1979. The NRC environmental statement for Charlestown, Rhode Island,[9] assumes a base capacity factor of 60 percent over the thirty-year life for both nuclear and fossil-fuel plants, with alternatives of 50 or 70 percent. There is no mention of senescence. However, NRC staff elsewhere does make a senescence assumption derived from the Bown-Snyder sequence.[10]

New England Power Company (NEPCO), 1979. For the proposed Charlestown, Rhode Island, nuclear-power plant, NEPCO in October 1977 assumed a capacity-factor rise from the first year to a plateau for the sixth year through the thirty-year life, as follows:

Year	CF (%)
1	59.2
2	60.9
3	66.8
4	71.0
5	71.0
6–30	76.2
30-yr average	74.5
40-yr average	74.9
28-yr average	74.3

NEPCO's sequence, assuming zero senescence, is typical of the literature in the industry.[11]

A.D. Little/Stoller Study, 1975. The ADL/Stoller study was made for NEPCO and served as the basis for NEPCO's decision to build the Charlestown nuclear-power plant.[12] The study assumes a rise in capacity factor for the first four years and a leveling thereafter at 70 percent:

Year	Capacity Factor (%) Nuclear & Coal with FGD	Coal (& oil) without FGD
1	53	57
2	60	64
3	65	69
4–30	70	75

The study also assumes 62 percent from the fourth year as a low capacity factor for nuclear, and 78 percent as the high option, with 70 percent as the average.

The ERDA Study, 1976. The ERDA study, done by the MITRE and TRW companies, assumes a senescence factor that is not expressed precisely.[13] Coal and LWR generation are assumed to have similar lifetime capacity factors, which start at 50 percent, rise to 70 percent after two or three years, and "later steadily decreasing until the plant is retired."

In an appendix illustrating revenue requirements, the following senescence sequence is used for coal with FGD becoming commercial in 1981:

Year	CF (%)	Notes
1	60	
2	65	
3	70	The peak
4–18	69.7	Dropping steadily to 60%
19–30	58.6	Dropping more sharply, but steadily to 30%

It is not clear, however, whether this illustration is intended to represent both nuclear and coal with and without FGD.

AEC, 1974 and 1975. In its annual reveiw published in 1974, AEC gives its first analysis of the competitive economics of nuclear and coal units.[14] The review assumes a lifetime capacity factor of 80 percent for both, without any mention of senescence. The review in 1975 drops the capacity factor to 75 percent.

Adjustment to Case Studies

The original capacity factor assumptions for the four case studies, presented here, do not all reflect what coal and nuclear units can achieve in the real world. (The variability among the case studies is also of note.)

	Nuclear	Coal
AEC	75%	75%
ERDA	65	58
NRC	60	60
Exxon Hi-S with FGD	60	66
Exxon Low-S without FGD	60	70

U.S. nuclear units have experienced an average nameplate capacity factor of about 55 percent plus or minus a few percentage points in any one year (for example, 53 percent in 1973, and 55 percent in 1974 and 1975). The lifetime capacity factors for the thirteen U.S. nuclear units of at least 1,000 MW averaged 53.3 percent (nameplate) through 1981, with a range of 44.3 to 62.9 percent. Since nuclear operates as baseload like run-of-stream hydro, nuclear-capacity factors must be taken as the maximum achievable under actual long-run conditions. The seven large (greater than 1,000 MW) nuclear units with capacity factors over 55 percent (that is, the best large units) had an average lifetime-capacity factor (through 1981) of 58.6 percent, with a range of 55.2 to 62.9 percent.

A 55-percent capacity-factor adjustment is used for nuclear units in the base case of the four case studies. The sensitivity analysis assumes 50-percent and 60-percent capacity factors for nuclear. The 55-percent base-case lifetime capacity factor may be high given the indications of retrogressive learning (see chapter 4) and the fact that senescence has not been taken into account. The time pattern of senescence and the rate of fall-off are sufficiently indeterminate at this point that a reliable sequence is not feasible.

For coal, evidence indicates that capacity factors of 70 percent and even 75 percent are within the range of experience. For 1972 FPC data for the twenty-four most efficient coal-burning units in the United States utility industry (units using under 9,100 Btu/kwh) showed an average capacity factor of 76 percent. For 1974, the eighteen most efficient units averaged 77 percent. Since fossil-fuel units are load following, with dispatchers giving preference to run-of-stream hydro, then hydro storage at peaks, followed by nuclear units, the capacity factors are fairly high. After 1974 fossil-fuel capacity factors begin to be distorted by the expansion in nuclear units.

Nuclear capacity factors are properly compared, because of load following, the equivalent-availability factors (EAFs) for coal. EAF is the ratio of a unit's available hours, reduced by partial outages, to period hours. For 164 U.S. coal units of 400 MW or greater, the EAF averaged 68.01 percent in 1979 and 69.45 percent in 1980 (National Electric Reliability Council, *Ten Year Review, 1971–1980, Report on Equipment Availability*).

The validity of a 70 percent capacity factor for coal is confirmed by the four, 1,300-MW coal units of American Electric Power Company (AEP). The four units, for which OAFs are given in table 6–9, have twenty-four unit-years of combined experience through February 1982. While EAFs are the preferred measure, they are not publicly available. The AEP stated in discussions (July 1982) that the EAFs run six to seven percentage points lower than these OAFs. The AEP capacity factors reflect load following as well as another reduction which is temporary: the long-term coal contracts

Table 6-9
Performance Data for American Electric Power's 1,300-MW Coal Units

	Annual								Peak[a]							
	Operating Availability (%)				Capacity Factor (%)				Operating Availability (%)				Capacity Factor (%)			
Year	M	G-1	G-2	A-3	M	G-1	G-2	A-3	M	G-1	G-2	A-3	M	G-1	G-2	A-3
1974	—	98	—	83	—	87	—	66	—	—	—	100	—	—	—	83
1975	—	84	99	83	—	65	78	67	—	65	—	98	—	56	—	85
1976	—	87	87	74	—	70	71	61	—	99	85	98	—	81	71	87
1977	—	86	83	86	—	69	65	76	—	100	92	88	—	86	76	84
1978	—	88	86	88	—	70	67	74	—	96	94	89	—	78	78	65
1979	—	82	88	84	—	69	74	76	—	100	100	93	—	87	89	90
1980	96	84	86	84	79	68	69	70	—	76	100	99	—	69	92	93
1981	97	84	86	80	79	53	61	60	94	100	88	89	83	74	62	68
1982	—	—	—	—	—	—	—	—	90	88	100	100	69	59	69	74

Source: Mechanical Engineering Division of American Electric Power in New York City.

Note: Mountaineer = M, Gavin = G, Amos = A. Analysis based on nameplate ratings. Commerciality dates are: Mountaineer, September 15, 1980; Gavin-1, October 20, 1974; Gavin-2, July 6, 1975; Amos-3, October 22, 1973.

[a]January and February are used as proxy for the peak winter period.

supplying the 1,300-MW units are more recent and substantially higher priced than contracts for older and smaller coal units. Until these older contracts expire, it is frequently more economic to run the older, though smaller and perhaps less efficient units. Upon expiration of the older contracts, the capacity factors will increase for the 1,300-MW units, ceteris paribus. Statistics for large coal units cannot be ascertained as industrywide unit data for coal are not publicly available. Scale problems, however, are not as significant for coal as for nuclear units.

For the four units combined (counting only full years), the lifetime OAF through 1981 averaged 85 percent. The newest unit, Mountaineer (1980), had a 97 percent OAF for its first full year (1981). The average lifetime capacity factor for the four units is 68 percent, with Mountaineer again the highest at 79 percent. For 1981, OAF for the four units averaged 87 percent and their capacity factors averaged 63 percent.

Operating availability at annual peaks is an essential characteristic of reliability. AEP has had a winter peak with a second summer peak. January and February have been used as a proxy for the peak winter period. For the twenty-six winter peaks of the four units (through 1982), OAFs averaged 93 percent compared with lifetime OAFs of 85 percent. Capacity factor at peaks averaged 77 percent for the four units, compared with 68 percent for lifetime. At the 1982 peak, OAF averaged 95 percent and capacity factors averaged 68 percent. The AEP states that with appropriate maintenance, the preceding OAFs should prevail (approximately) for the thirty-three year economic life of the units (conversations with AEP staff in April 1982). It is important to note that all of the preceding OAFs would still be well above 70 percent when adjusted downwards by 6 percentage points to approximate EAFs.

The annual capacity-factor adjustment for coal units in the four case studies is 70 percent for the base case. Based on the data presented previously, 70 percent is considered to be conservatively low. Sensitivity analysis is also performed assuming 65 percent and 75 percent capacity factors for coal.

Notes

1. To avoid confusion, *sufflation* has been used for technological inflation reserving *inflation* for pure price-level rises.

2. Charles Komanoff in, *Power Plant Cost Escalation* (published by Komanoff Energy Associates, New York City, 1981), estimates nuclear and coal capital costs for 1988 using regression analysis. While his study and this study are very different (for example, Komanoff compares very small coal and large nuclear units), it is interesting to note that the ratio of coal to nuclear capital costs ($/kw) which he derives (59 and 57 percent using mixed and constant dollars, respectively) is similar to the ratios derived in the four

case studies after adjusting nuclear for technological sufflation (AEC = 61 percent; ERDA = 63 percent; NRC = 58 percent; Exxon High-Sulfur Coal = 56 percent).

3. See chapter 2 for a more detailed discussion of the difficulties.

4. See chapter 2, note 1, for sources of European material.

5. Nuclear-unit lives of more than thirty years have been considered. NRC in its Charlestown study mentioned nuclear-unit lives of forty years at three points. Some utilities, including NEPCO, have mentioned forty years for nuclear units, but this is an unjustifiable assumption at this time.

6. If senescence were incorporated, the mills/kwh increase would be lower. (Senescence is discussed later in this chapter.) However, given the existence of other offsetting factors, the straight-line assumption is generally a fair approximation.

7. Conversation with Dr. Trevor Broom, Director of Operations, British Central Electricity Generating Board (BCEGB).

8. Robert Bown and Artha J. Snyder, "Total Energy, Electric Energy, and Nuclear Power Projections," Energy Research and Development Administration (ERDA), February 1975, mimeograph.

9. U.S. Nuclear Regulatory Commission, *Draft Environmental Statement, Charlestown, RI Nuclear Power Plant,* NUREG–0529, May 1979.

10. Conversation with Darryl A. Nash, July 1980. The sequence is found in "Coal and Nuclear: A Comparison of the Cost of Generating Baseload Electricity by Region," U.S. Nuclear Regulatory Commission, December 1978.

11. U.S. Nuclear Regulatory Commission, *Draft Environmental Statement, Charlestown, RI Nuclear Power Plant,* NUREG–0529, May 1979.

12. Arthur D. Little, Inc./S.M. Stoller Corp., "Economic Comparison of Baseload Generation Alternatives for New England Electric," Cambridge, Mass., 1975, p. 71.

13. ERDA "Comparing New Technologies for the Electric Utilities, Draft Final Report" December 9, 1976, 76–141, p. 102–103 and A-5.

14. U.S. Atomic Energy Commission, *The Nuclear Industry 1974,* WASH 1174-74, p. 20, and *The Nuclear Industry 1973,* WASH 1174-73, p. 15.

7 Fuel and Operating- and-Maintenance Costs

Two major operating costs of power generation are addressed in this chapter:

Fuel costs
> Btu per pound of coal
> Heat rate for coal
> Fuel prices

Operating and Maintenance Costs (O&M)

Each of these factors is discussed separately and, except for fuel prices, is developed into a quantified adjustment for the case studies.

Btu Per Pound of Coal

The Btu per pound of coal is a conventional measure that was used in the AEC and NRC case studies. ERDA and Exxon used cents per million Btu, which bypasses the pound measure. The AEC's assumption of 10,900 Btu/lb is an error; it is actually the average of all coals used by utilities in 1974, including low-sulfur lignite.[1] Steaming coal has about 12,000 Btu/lb, which is the NRC assumption.

Many plants show coal values at 12,500 Btu and higher. A trend toward better cleaning and preparation of coal at the mine is underway and will result in higher Btu/lb at the generating plant. With the increased freight rates for moving coal and the reduced handling costs at the generating plant, heating values of coal will inevitably surpass 12,000 Btu by the time plants now ordered go on line.[2] After the oil embargo of the fall and winter of 1973–1974, the squeeze on coal supply resulted in a deterioration of delivered quality and Btu/lb. Now, however, the demand for better coal and greater competition among coal suppliers should result in improved quality. Accordingly, a value of 12,000 Btu/lb for midwestern-bituminous coal has been used for the AEC adjustment. (No fuel-price adjustment is necessary as the AEC's assumed fuel price is for midwestern-bituminous coal.)

Heat Rate for Coal

Heat rate is expressed in Btu of coal per kwh. The four cases made the following assumptions:

AEC:	9,500 Btu/kwh for high-sulfur coal
ERDA:	9,751 Btu/kwh for high-sulfur coal
NRC:	10,000 Btu/kwh for high-sulfur coal
EXXON:	9,420 Btu/kwh for high-sulfur coal
	9,050 Btu/kwh for low-sulfur coal

In the Federal Energy Regulatory Commission's (FERC)[3] annual *Statistics of Steam-Electric Plants,* the "best" generating units are defined as having heat rates of less than 9,100 Btu in 1972, and 9,200 Btu in 1974.[4] In 1972 some twenty-four coal units using bituminous had heat rates under 9,100 Btu, with the lowest unit at 8,846 Btu. In 1977 eleven units had rates under 9,200 Btu, the lowest being 8,842 Btu. Without FGD, therefore, a 9,000-Btu heat rate is reasonably attainable for new, well-designed and well-built units using coal cleaned and prepared to state-of-the-art standards. FGD uses additional energy equal to 1–2 percent of total generation. At 2 percent, the 9,000-Btu without FGD becomes 9,180 Btu. The adjustment used in the four case studies is 9,200 Btu/kwh for high-sulfur coal. Exxon's low-sulfur-coal assumption did not require adjusting.

Fuel Prices

The fuel prices for coal and nuclear in the four case studies are not adjusted. The complexity of the methodologies, sometimes compounded by insufficient information, and the uncertainty of fuel-price projections in general, made a reasonable fuel-price adjustment infeasible.

Fuel-cost escalation in the next ten years is likely to be faster for nuclear than for coal. In general, domestic and international competition among suppliers of coal is greater than within the uranium and OPEC cartels. The initial price surge for coal after the oil embargo of 1973–1974 peaked in 1975. It was followed by expanded supply that continues to grow on a worldwide basis. The result has been not only a leveling off of coal prices in constant dollars but a drop. Because of the competitive situation in coal production, "fundamentally the price of coal is determined by its production cost."[5] Further, compared to coal, uranium reserves are fairly limited without the breeder, which is not a realistic option for the next decade. Because AEC, ERDA, and NRC assumed identical coal and uranium fuel cost inflation these studies underestimate nuclear's fuel costs.

Operating-and-Maintenance Costs

Operating and maintenance costs (O&M) illustrate well the design/performance gap of nuclear power. Nuclear power's O&M costs at design are always assumed to be below those for coal. All four case studies originally made this assumption. In actuality, the reverse relationship holds.

Comparing large (over 800 MW) units in 1978 (tables 7–1 and 7–2), O&M for nuclear averaged 97 percent more than for coal. (The average capacity factors for the twenty-three nuclear units and for the fifteen coal units were 55 percent and 53 percent respectively.) Raising the coal units' average O&M by 20 percent for FGD makes nuclear O&M 65 percent higher than coal O&M.

Table 7–1
O&M for Large Nuclear-Power Units (over 800 MW), 1978

Unit(s)	Year First Unit Commercial	Nameplate MW	CF (%)	O&M Million $	$/kw	Mills/kwh
Farley	1977	888	76	12.21	13.75	2.06
Browns Ferry (3)	1974	3,456	52	45.92	13.29	2.90
Arkansas-1	1974	902	66	12.12	13.44	2.33
Rancho Seco	1974	928	61	11.83	12.74	2.37
Millstone-2	1975	910	56	22.29	24.49	4.94
St. Lucie	1976	850	67	15.81	18.60	3.18
Crystal River	1977	890	33	15.61	17.54	6.01
Zion (2)	1973	2,196	70	20.38	9.28	1.50
Maine Yankee	1967	864	71	10.82	12.52	2.01
Palisades	1970	812	37	15.39	18.95	5.91
Cooper	1974	836	67	8.31	9.94	1.69
Salem-1	1977	1,170	44	22.31	19.07	4.93
Fitzpatrick	1975	883	54	19.04	21.56	4.54
Indian Point-3	1976	1,068	58	23.32	21.84	4.27
Davis-Besse	1977	906	31	14.10	15.56	5.76
Trojan	1976	1,216	16	15.20	12.50	9.13
Beaver Valley	1976	923	30	22.68	24.57	9.29
TMI-1	1974	871	74	17.95	20.61	3.17
Calvert Cliffs (2)	1975	1,828	62	26.00	14.22	2.63
Average (23 units)			55		15.85	3.85

Source: Adapted from U.S. Department of Energy, *Steam Electric Plant Construstion Cost and Annual Production Expenses,* 1978.

Table 7–2
O&M for Large Coal Power Units (over 800 MW), 1978

Unit(s)	Year First Unit Commercial	MW Nameplate	CF (%)	Mills/kwh
Amos (3)	1971	2,933	66	1.11
Bowen (4)	1971	3,499	45	2.52
Cumberland (2)	1973	2,600	46	2.52
Bull Run	1967	950	55	2.25
Gavin (2)	1974	2,600	64	0.83
Paradise (3)	1963	2,558	47	2.28
Average (15 units)			53	1.95

Source: Adapted from U.S. Department of Energy, *Steam-Electric Plant Construction Cost and Annual Production Expenses,* 1978.

Nuclear's design/performance gap for O&M results from the techno-logical problems discussed in part I. Technological problems affect average O&M in two ways—total O&M increases and the number of kwh (that is, capacity factor) over which total costs are averaged decreases. Thus capac-ity-factor and O&M fluctuations should go hand in hand for nuclear if O&M costs are fully reflected in utility accounts. Low capacity factors for fossil-fuel units, in contrast, should not generally be accompanied by higher O&M. Their lower capacity factors are generally due to load following, not technological problems (see chapter 6). Thus while the number of kwh decreases, total O&M also decreases, resulting in little change in average O&M.

In the four case studies, the nuclear O&M adjustment is based on the 1978 average O&M for large nuclear units (with a 55-percent average capac-ity factor) in table 7–1. For AEC, ERDA, and Exxon, the 1978 average O&M for nuclear was escalated or deflated (using the individual case study's escalation rate) to the appropriate year. NRC's methodology made direct use of the 1978 average O&M inappropriate. Instead, NRC's nuclear O&M was adjusted so that the ratio of coal to nuclear O&M equaled Exxon's adjusted ratio for the high-sulfur-coal scenario.

Coal O&M was not adjusted in the four case studies. According to the NRC's computer model for projecting coal O&M, capacity-factor changes affect average O&M very little because the numerator (total O&M) and the denominator (kwh) move in the same direction. For example, a capacity-factor increase from 60 percent to 70 percent decreases O&M by only 1.2 percent; a 75-percent capacity factor decreases O&M by only 2.5 percent (over the 60-percent case).[6] These changes are too small to warrant an adjustment.

Notes

1. This is exactly the Btu/lb shown for bituminous and lignite coals combined as used by electric utilities in 1974. See U.S. Department of Energy, *Monthly Energy Review* (Washington, D.C.: Government Printing Office, various issues), back cover showing conversion rates.

2. See, for example, "Coal-Cleaning Test Facility," *EPRI Journal,* November 1980, p. 40.

3. Formerly the Federal Power Commission (FPC).

4. FERC/FPC, *Steam-Electric Plant Construction Cost and Annual Production Expense* (Washington, D.C.: Government Printing Office).

5. President's Commission on Coal, *Coal Data Book,* 1980, p. 96.

6. The NRC model attributes approximately 45 percent of coal's O&M costs to ash-waste disposal. Ash, however, can be a source of revenue. Ash is now being sold for agricultural use and for the production of cement and concrete. The Environmental Protection Agency estimates that, of the potential 60 to 65 million tons of fly ash in 1985, approximately 10 to 20 million tons could be used in cement and concrete (*Electrical Week,* McGraw-Hill, January 19, 1981, p. 11).

8 Reliability-Related Costs

The economic implications of the design/performance gap for nuclear power have been partially incorporated into the four case studies via the capital-cost and O&M adjustments discussed in the previous two chapters. Three costs for nuclear have not been included because their quantification was problematic. The three costs, discussed separately here, include the following:

The Yoyo Factor

Replacement power

Major repairs

The Yoyo Factor

The annual charts of daily capacity factors for nuclear power plants, known as *histograms,* indicate that capacity factors rise and fall like a yoyo with considerable frequency. This fluctuation can be more important than the capacity factor itself. The Yoyo Factor is vital in any economic comparison of nuclear and fossil-fuel power, but it has been virtually ignored in the literature. Reliability is related to the Yoyo Factor but, in the literature, reliability is generally and erroneously equated solely with capacity factor.

As discussed in chapter 6, capacity factor is the percent of nameplate rating that is actually achieved over a specified time period. In calendar 1979, for example, U.S. nuclear-power units had a nameplate capacity factor of 55.7 percent. Reliability includes capacity factor but, more important, also includes the degree to which on-line availability can be predicted and controlled. Thus two plants with 55-percent capacity factors are viewed entirely differently if, for one, the availability can be controlled at peak times while, for the other, availability is only poorly predictable. Histograms of nuclear-power units illustrate that shutdowns occur frequently during peaks and at other times when utilities clearly would not so choose voluntarily. Therefore, there is a strong element of stochastic variation.

The Yoyo Factor in nuclear power plants is in good part due to their high sensitivity to safety. Automatic shutdowns (scrams) are frequent and may be due to faults in metering, sensing, and controls, as well as to breakdowns in primary- or secondary-circuit equipment. For the same reasons, manual shutdowns are also frequent.

Nuclear and Fossil-Fuel Comparisons

The nuclear power industry is unique in publishing monthly and annual histograms for every U.S. unit, and annual histograms for noncommunist countries. Such histograms, permitting visualization of the Yoyo Factor, are not published for fossil-fuel units in the United States or Europe.

Capacity-factor histograms for fossil-fuel and nuclear units, even if available, would not be directly comparable. Since nuclear units should be operated whenever available, in principle their capacity factor would measure equivalent availability. Fossil-fuel units, being load following, would have a capacity factor lower than equivalent availability.[1] Therefore, the correct comparison is between nuclear capacity factor and fossil-fuel equivalent availability histograms.

It appears that fossil-fuel units have an availability histogram with much less of a yoyo pattern than nuclear histograms. Dr. Schoch, head of the Mannheim plant, must sell his power wholesale competitively.[2] He has stated in conversations that he could not operate with the nuclear histogram. He must have 90-percent availability at the winter peak, with three-hour overload capability. The Yoyo Factor was a major reason for his not recommending purchase of a nuclear plant.

Other factors that indicate less of a Yoyo Factor for coal than for nuclear include:

Fossil fuel is a more mature technology.

Fossil fuel is a safer technology requiring much less provision for automatic scramming.

Fossil fuel is a less complex technology, much more within human capability to operate and maintain.

Fossil fuel has involved much less breaching of scale on the larger units (see part I).

Economic Impacts

The Yoyo Factor is another manifestation of the design/performance gap of nuclear power. It is perhaps as important to a valid economic comparison of nuclear and fossil fuel as capacity factor. As with other aspects of the design/performance gap, the Yoyo Factor has not been adequately recognized.

The ERDA case study, for example, ranks existing and prospective technologies by complexity in the following order: coal without FGD, solar, ocean thermal, atmospheric fluidized bed (coal), coal with FGD, hydrothermal, LWR, open cycle (gas turbine/steam with integrated gasifier), the liquid-metal fast breeder (LMFBR), fusion, and solar satelites. Nevertheless, the study assumes equal capacity factors and reliability for coal and the LWR: "In effect, this assumption involves treating all technologies as having comparable reliability regardless of their inherent complexity" (p. 103). The NRC case study goes further: "The New England nuclear plant [at Charlestown] will increase the reliability of the region's power supply" (p. 10–7). It gives no explanation except possibly that nuclear fuel is a more reliable supply than imported oil.

The economic impacts of the Yoyo Factor are twofold—the costs of replacement power and of reserve capacity substantially greater than would be required for a comparable fossil-fuel unit. Neither of these costs, crucial as they are to economic comparisons, has been incorporated into the case studies. The difficulties with incorporating replacement power are discussed in the next section. Quantifying reserve-capacity costs requires a study unto itself, which was not feasible for this analysis. The size and type of reserve capacity are determined by a utility's current capacity mix, projected load growth, and interconnections with other utilities.

Replacement Power

As the largest nuclear-power units have come on-line and as prolonged unplanned outages have occurred with some frequency (the Yoyo Factor), it has become clear in recent years that the cost of replacement power is a major component of nuclear-power costs. As stated in part I, every nuclear unit in operation as of 1982 was designed to operate at an 80-percent capacity factor or higher. Long-range matching of power supply and demand has been based on these capacity-factor assumptions. The 55-percent average capacity factor actually realized has created shortages of supply concentrated at plants with unplanned complete shutdowns. Examples include the two Browns Ferry units and the two TMI units.

This "distress" type of unplanned replacement power is very costly. It is obtained either from the utility's own units that are less efficient because of higher heat rates or from sources that charge the full costs of their marginal and less efficient units. Distress power is distinct from the much cheaper power purchased by utilities on long-term commitments for government hydro, or joint ownership of hydro and fossil-fuel generating units. The latter is termed *purchased power* as distinct from *replacement power*.

The national average cost of purchased power includes all types: distress, hydro, joint ownership, gas, oil, coal, and nuclear. As indicated in table 8–1, the cost rose from 8 to 23 mills between 1973 and 1979. The increase of 178 percent is due mainly to higher fossil-fuel costs. General inflation, as measured by the GNP price deflator, increased by only 56 percent.

A good illustration of distress-power pricing is the shutdown of the two TMI nuclear-power units on March 28, 1979, which continues into 1982. The shutdown had the following effect on purchased power for Metropolitan Edison Company:

Cost of Purchased Power, ¢/kwh

Year	Met. Ed.	U.S.
1978	1.42¢	2.02¢
1979	4.79	2.346
Increase	237%	16%

Source: Adapted from U.S. Department of Energy, *Statistics of Privately Owned Electric Utilities in the U.S.,* 1979.

Replacement-power costs for nuclear are difficult to quantify because of the wide range of possible prices. Estimation is further complicated by the likelihood that utilities would not indefinitely make up for capacity-factor shortfalls with replacement (that is, distress) power. Once the chronicity of the shortfalls was perceived, arrangements would be made to build additional capacity or to purchase power under long-term contracts. Thus the

Table 8–1
Purchased Power in the U.S. Electric-Utility Industry, 1973–1979

Year	Kwh (Billions)	Dollars (Millions)	Mills/Kwh	Mills/Kwh Index	GNP Price Deflator
1973	232	1,955	8.43	100	100
1974	227	2,726	12.01	142	110
1975	230	3,239	14.06	167	120
1976	251	4,027	16.04	190	127
1977	237	4,653	19.63	233	134
1978	261	5,275	20.21	240	144
1979	257	6,034	23.46	278	156

Source: Adapted from FPC/U.S. Department of Energy, *Statistics of Privately Owned Electric Utilities of the U.S.* Price deflator from *Statistical Abstract* and U.S. Department of Commerce's *Survey of Current Business.*

cost of the shortfalls would constitute a mixture of distress-priced and non-distress-priced kwh. For example, General Public Utilities (Metropolitan Edison's parent company) made an agreement in June 1982 (three years after the TMI accident) to purchase replacement power from Detroit Edison Company for eight and one-half years at 4.7¢/kwh. The 4.7¢ compares with the distressed-priced replacement power in the previous three years of 6.5¢ (Bureau of National Affairs, *Energy Users Report,* July 1, 1982). The 6.5¢ represents a substantial increase over the cost for 1979 of 4.79¢.

The complex and uncertain nature of pricing nuclear capacity-factor shortfalls prevented the development of a replacement-power adjustment for the four case studies. Nonetheless, replacement power is a real and substantial cost of nuclear power that deserves further attention.

Major Repairs

Low-capacity factors for nuclear not only reduce the number of kwh over which total costs are averaged but may reflect major technological problems requiring costly repairs. Major repairs have become an important cost category for nuclear. Two problems that may well require major repairs (not to mention pose serious safety hazards) for a significant portion of the industry were discussed in part I—embrittlement in reactor vessels (eight units) and steam-generator tube leaks (forty units). The latter includes the Ginna plant near Rochester, New York, where, in January 1982 a steam-generator tube ruptured releasing radioactive steam into the air.

Another case in point is Florida Power and Light's (FPL) units, Turkey Point-3 and -4. The two units went commercial in September 1972 and March 1973, respectively. Beginning in August 1974, FPL discovered corrosion and other related problems with the steam-generator tubes. The problems became progressively worse. By April 1975 for unit 3 and October 1975 for unit 4, the steam generators were "totally unfit for their intended purposes, and reasonable efforts to render them fit or suitable (had) been unsuccessful."[3]

In 1978, FPL filed suit against Westinghouse Electric Corporation (WEC) who had designed and manufactured the nuclear steam-supply systems for FPL. FPL's complaint stated that the systems were "negligently designed and manufactured. . . . Improper materials were not corrosion resistant." The design and manufacture were said to "facilitate corrosion." FPL claimed that operating instructions "negligently specified the introduction of chemicals and substances into the liquid transported around the tubes, which facilitated the corrosion of the tubes and support plates." FPL stated that WEC failed to warn of these potential problems even though they had previously occurred. The cost of replacing the steam-generator systems approximated the original cost of Turkey Point-3 and -4.

In an interesting parallel to the Bechtel Theorem (see part I), WEC denied responsibility for the problems. In its reply, WEC held that express warranties had expired before the defects became evident, that implied warranties had been excluded from the contract, and that the statute of limitations had expired before the FPL suit. WEC stated that FPL was "itself guilty of contributory negligence in its supervision, control, operation, and maintenance . . . which was the legal cause of all the losses . . . complained of." This and breach of contract of third parties were the "sole legal cause" of losses. WEC claimed that FPL could not sue because it had

> reviewed and had its contractor Bechtel review Defendant's design for the subject equipment and accepted that design as being the state of the art in the nuclear power industry at the time of contracting and by reason of this review assumed the responsibility for the type of problems complained of.

The Three Mile Island-2 (TMI) cleanup exemplifies an extreme case of major repairs. A recent estimate (*Life,* May 1982) predicts that "the cleanup will take 20 years at the rate it's going and cost the owner, General Public Utilities, at least $1.1 billion—more than the plant cost to build" (p. 42). Replacement power, necessitated by the TMI shutdown, accompanies most major repairs.

The accounting treatment for major nuclear repairs is difficult to determine. These repairs, which can run to hundreds of millions of dollars, such as for replacement of steam generators, usually are not included in O&M. The costs are generally incorporated into a special reserve fund and amortized over five to ten years. They are more properly O&M rather than a "capital improvement." Similarly, large O&M performed under contract may find its way into accounts other than O&M for the nuclear unit.

Notes

1. See discussion of the "hydro analog" in chapter 6.
2. See chapter 2, note 1, for sources of European information.
3. All quotes are from *Florida P&L* v. *Westinghouse Electric Corporation,* Complaint for Damages, U.S. District Court, Southern District of Florida, Case No. 78–1896–Civ–CA, May 10, 1978, and Westinghouse's "Answer" of May 16, 1979.

9 Waste Disposal and Decommissioning

Radioactive wastes (*radwaste*) are generated during the operation of nuclear power plants and, postoperationally, as products of decommissioning. The technological and economic principles of disposal are the same for both. Radwaste disposal encapsulates and extends the problems of nuclear power generally—the great technological uncertainties and the resulting economic uncertainties.

Radwaste disposal has three points of uncertainty. First, has the industry realistically defined the problem? For example, has the volume of materials for long-term storage been correctly estimated? The quantity of material has been increased by steam-flow vibration, corrosion, steam-generator replacement, and long-lived irradiation of inner and outer containment vessels. Further, the problem definition must encompass unexpected materials and disposal procedures from, for example, the partial meltdown at TMI. Second, are the strategies technologically sound, and do they involve acceptable safety risks? Third, have the costs for the strategies been realistically assessed? The record of economic prediction in the nuclear power industry thus far does not provide a positive precedent.

The preceding points of uncertainty all assume that a solution is possible but that it may be deficient technologically, safety-wise and economically. A more fundamental question about radwaste disposal must be asked: Is the problem inherently unsolvable? According to the U.S. Atomic Energy Commission, "Commercial high level wastes . . . have such a high radioactive content of long-lived isotopes that they require long-term storage in isolation and under essentially perpetual surveillance at the storage sites."[1] If a perpetual priesthood is unrealistic, is a solution equally unrealistic?

The uncertainties of radwaste disposal, discussed below, precluded the incorporation of a waste-disposal and decommissioning adjustment into the four case studies. The original studies included zero or insignificant amounts for both. Following the next section on general background, the state-of-the-art costs for radwaste disposal are discussed.

Background

Radwaste has three levels: high-level waste (HLW), transuranic, and low-level waste (LLW).[2] HLW originates in fuel rods. It contains nuclides with

91

half-lives of "one year to millions of years."[3] Transuranic wastes are more highly radioactive than LLW, including significant amounts of long-lived alpha emitters such as plutonium. These components are included in long-term waste disposal. Examples include filters, rags, paper, plastics, tools, contaminated equipment, and fuel hulls. LLW contains isotopes of uranium whose lives are so short that protected shallow-land burial is deemed safe by NRC. This category is not considered here. Examples include residues from chemical processing, building rubble, metal, wood, and fabric scrap; glassware; paper and plastic; plant waste; sludges and acids; and slightly contaminated equipment or tools.

During the operating lives of nuclear power plants, radwaste is generated from spent fuel rods; from replacements of steam generators, tubing, piping, valving, and pumps; from cleaning materials; from the offal generated by cleaning such as from turbines of BWRs; and from clothing. Spent fuel rods must be left onsite at least ninety days but more usually as much as ten years if storage pools are available. Pending permanent storage solutions, the NRC and DOE are exploring the establishment of intermediate-term storage pools offsite. Otherwise, power plants may be forced to shut down for lack of onsite storage. Similar considerations apply to replacement of steam generators and other highly irradiated plant components.

During decommissioning, radwaste arises from fuel; from removable equipment such as steam generators, tubing, and so on; and from permanent structures such as steel and concrete containment vessels and the building proper. Three types of decommissioning are defined by NRC: decontamination (decon), safe storage (safstor), and entombment (entomb).[4] *Decon,* which formerly was called "dismantlement," consists of immediate total removal of radioactive materials to permanent storage, thus permitting unrestricted use of the site. "Immediate," per NRC, means four years. *Safstor,* formerly known as "layaway," involves the removal of items that can be quickly transported offsite, secure storage of the remaining items onsite for up to a hundred years (during which time radioactivity is substantially reduced), followed by decon and restoration of the site to unrestricted use. *Entomb* entails encasement of the entire power structure with "surveillance in perpetuity." If, however, the internals of the pressure vessel and nickel-59 and niobium-94 are removed to a repository, surveillance could be terminated "within several hundred years."

NRC prefers immediate decon, rejects entomb, with safstor somewhere in between. The Federal Energy Regulatory Commission (FERC), in its first rate case considering decommissioning costs (1980), preferred entomb because it would put the least burden on rate payers before decommissioning.[5] In 1982 FERC staff takes the NRC position of preferring decon.

Table 9-1 indicates for the various types of decommissioning, the required security, staffing, environmental monitoring, and surveillance. A number of cases include no onsite staffing, and the degree of surveillance varies across cases.

The components in table 9-1 for spent-fuel disposal in 1980 dollars, as estimated in 1982, are as follows in mills/kwh:

Onsite storage	0.24
Transport	0.06
Disposal fee	0.99
Total	1.29

The 0.99 mill is an assumed one-time charge by the U.S. government to utilities for perpetual storage.

A brief note on the physical problems of decommissioning helps to visualize the process. In a PWR, steam generators, primary-circuit piping, the reactor proper, waste handling, and the concrete biological shield are major radiation problems. The BWR adds the steam turbine, through which radioactive steam passes. Leakages from radioactive into normally nonradioactive circuits augment problem equipment. Remotely operated plasma arc torches may be used for cutting 1- to 1½-in. stainless steel under water and 3½-in. carbon steel in air.[6]

The NRC's "regulatory design philosophy" is "to maintain radiation exposure As Low as is Reasonably Achievable" (ALARA). Taking this phrase at face value, storage need not provide adequate protection to the

Table 9-1
NRC Decommissioning Modes: Nature of Onsite Costs

Decommissioning Type	Security	Continuing-Care Staff	Environmental Monitoring	Surveillance
1. Decontamination	0	0	0	0
2. Safe storage				
a. Custodial	Continuous	Some	Continuous	Continuous
b. Passive, mothball	Remote alarms	Optional	Routine periodic	Periodic
c. Temporary entombment	Barriers, fencing, & posting	0	Infrequent	Infrequent
3. Entombment	Barriers	0	Infrequent	Infrequent

Source: Adapted from Nuclear Regulatory Commission, *Draft Generic Environmental Impact Statement on Decommissioning of Nuclear Facilities* NUREG–0586, January 1981, p. 2–6.

public. The standard is not an absolute value but a relative one, dependent on state of the art and economics. The ALARA and perpetual surveillance concepts indicate that the nuclear power industry has been operating, and continues to operate and expand capacity, without knowing how to manage its lethalities.

Radwaste-Disposal Economics

The economics of radwaste disposal has two points of focus: the cost of disposal procedures themselves and the financing costs. Each is addressed separately here.

Cost of Waste Disposal

The costs of waste disposal have three components: (1) storage of fuel rods onsite, normally while the generating unit is operating, the costs of which are relatively easy to calculate; (2) offsite interim rod storage, awaiting either a permanent storage site or reprocessing (if the present policy of non-reprocessing were to change), which would generate HLW for permanent storage; and (3) the ultimate problem of long-term "perpetual" storage.

Apart from the uncertainties inherent in waste disposal, differences in assumptions and methodologies can produce big differences in costs. Four assumptions—generating unit life, capacity factor, inflation and discount rates, and catastrophe—are discussed briefly here.

The NRC, based on work by Battelle Pacific Northwest Laboratories (PNL), uses a forty-year unit life in its decommissioning study (though it generally uses a thirty-year life in economic studies). Using a more realistic number of twenty-five years (see chapter 6), the cost per kwh increases 60 percent.

Capacity-factor assumptions vary widely in the literature. Reducing the most commonly used 70 percent to the more realistic 55 percent (see chapter 6), increases waste-disposal costs per kwh by 36 percent.

Changing inflation and discount rates by a few percentage points in either direction can change the cost of waste disposal by tens of millions of dollars. A further problem is that the NRC case study, for example, uses a 5-percent inflation rate and a 10-percent discount rate, a difference that is not explained.

TMI-2 has established definitively that any scenario for decommissioning and waste disposal must include provision for a partial or full meltdown. Assuming that TMI-2 will cost a minimum of $1 billion to dismantle, the early industry estimate of approximately $50 million increases by a factor of 20.

Cost of Spent-Fuel Disposal. DOE's fee estimates for permanent spent-fuel disposal illustrate that, as knowledge increases, so do costs. Between April 1981 and April 1982, the fee estimates rose 61 percent (from $217.80 to $350.00), versus an inflation rate of 9 percent. The increase due to better information alone, that is, without inflation, is 47 percent. The revisions to spent-fuel-disposal costs, which are the same for each year from 1985 to 2020, are shown in table 9-2. To illustrate the uncertainties of estimation, the $350/kg fee compares with approximately $260 estimated by a different branch of DOE.[7]

Decommissioning Costs. The absence of a site-specific permanent waste-disposal solution, as distinguished from a purely theoretical solution of scientists and engineers, is the chief obstacle to new nuclear units. It has moved numerous states to ban new plants until a real solution is reached. In Germany, provincial courts have banned the completion of units under construction until a solution is achieved. In its *Draft Generic Environmental Impact Statement* of January 1981 on decommissioning, the NRC posits that the *statement* "is required because the regulatory changes that might result from the reevaluation of decommissioning policy may be a major NRC action affecting the quality of the human environment."[8] A good starting point for decommissioning-cost estimates is the *Generic Statement.*

Decon, or immediate decontamination and decommissioning of the site to long-term storage, is NRC's preferred solution. Decon requires planning during the last two years of a unit's operation, plus four years of disman-

Table 9-2
Change in DOE/EIA's Spent-Fuel-Disposal Estimates for 1985-2020

	Estimate Date			
	April 1981		April 1982	*Percent Change*
Spent-Fuel Stage	*(1979 $)* *(1)*	*(1980 $)*[a] *(2)*	*(1980 $)* *(3)*	*(3) ÷ (2)*
1. Storage 10 years on-site, no reprocessing, $/kg/yr	6.50	7.08	7.20	1.7
2. Transport to permanent storage, $/kg	17.40	18.77	21.00	10.7
3. Disposal fee, $/kg	217.80	237.40	350.00	47.4

Source: The 1982 estimates are in U.S. Department of Energy (DOE)/Energy Information Administration (EIA), *U.S. Commercial Nuclear Power,* DOE/EIA-0315, March 1982, p. 31. The 1981 estimates are in the *1980 Annual Report to Congress* by DOE/EIA, vol. 3, Forecasts, p. 177.

[a]Adjusted to 1980 dollars using a 9-percent escalation as suggested by U.S. Department of Energy/Energy Information Administration.

tling. The costs in 1978 dollars for a PWR and a BWR, respectively, are given as $33 and $44 million. This estimate uses the 1,216-MW Trojan unit as a model.[9] For Haddam Neck/Connecticut Yankee, a 600-MW PWR, decon was estimated independently in a FERC proceeding at $57 million (also in 1978 dollars), compared with NRC's $33 million for a unit twice the size.[10] Decon for Three Mile Island-2, 961 MW, was estimated at $110 million (in 1979 dollars) by its owner, General Public Utilities Corporation.[11] TMI's estimate, adjusted to 1978 dollars by the change in the producer price index (90 percent) is $99 million.[12]

The preceding estimates, made by the utility industry, are subject to much skepticism by others in the field. For example, hearings were held before the Michigan Public Service Commission in 1981 on Consumers Power Company's decommissioning costs for the Big Rock (60 MW) and Palisades (812 MW) units. Peter Skinner of the New York State Attorney General's office estimated that decommissioning would cost $250 million (1978 dollars).[13] At least $526 million was estimated for removal of the units by the Michigan Association of Community Organizations for Reform Now.[14] At the 1977 congressional hearings, Skinner testified that decommissioning for an 1,150 MW unit would cost 24 percent ($1,359 million) of the original plant cost of $1.5 billion.[15] The Environmental Action Foundation testified that decommissioning and geological disposal costs were unknown, and "guesswork": "It is possible [that] the decommissioning costs could approach the cost of building a nuclear plant."[16]

NRC's generic decommissioning-cost estimates are, like other industry projections, based on idealized scientific and engineering assumptions. Two examples are the allowances for geologic disposal and for surveillance of safstor and entomb. In the $33 million for decon, "deep geologic disposal" accounts for only $850,000 (p. 4–4). The capital investment for the large volume of radioactive materials to be buried plus the cost of perpetual surveillance is unlikely to be so small, even in 1978 dollars. For the other two forms of decommissioning, surveillance is estimated at only $40,000 a year. In 1978 dollars, this would pay for one man year. It assumes only occasional onsite inspection and the use of remote television, sensors, and so on.

Financing Methods

The funding of long-term disposal of wastes is also a dilemma for regulatory agencies. What if a nuclear power company in some distant future cannot fund the various forms of decommissioning and walks away? The possibility cannot be ignored. Thus staffs serve up two scenarios:

Assume that the utilities will pay for decommissioning

Assume that the utilities will not pay for decommissioning

The latter scenario sets aside a fund under public control that will provide the necessary financing. The fund can be established as a lump sum in the construction cost that, with interest, will finance perpetual surveillance for whatever method of decommissioning is chosen. Alternatively, an amount can be set aside during each year of operation (sinking fund), which would finance decommissioning at the end of the unit's life.

Apart from the perpetuity problem, two limitations operate on any method—inflation and present value. If inflation is not incorporated in the set-asides, and even moderate inflation occurs, the shortfall could be substantial.

The technique of discounted present value can yield extraordinarily low costs per lifetime kwh for decommissioning. A $1-million decommissioning cost in 2012 has a present value in 1982 of $57,000 using the NRC's 10-percent discount rate. Each $1 million of protection and maintenance for the radioactive site in the year 2042 becomes $3,249 in 1982. Each $1 million in the ninetieth year after decommissioning (2102) becomes $11 in 1982. However, corporations and governments do not put money aside today in the huge sums needed for the nuclear power industry, in order to accumulate some cash value thirty or one hundred years in the future.

Given the long time horizons of radwaste disposal, both present value and sinking-fund methodologies are problematic. Moreover, any catastrophe—political, military, natural, or otherwise—that breaks the continuity of perpetual surveillance destroys the scenario.

Uncertainties

A waste-disposal and decommissioning adjustment has not been included in the four case studies because of the technological and economic uncertainties. The cost estimates just cited illustrate the problems. Additional cases are given below.

American Electric Power Company, referring to the two Cook nuclear units of its subsidiary, states that "Indiana and Michigan Power Company is studying alternative methods of decommissioning its nuclear power plant but cannot reasonably estimate at this time the future costs that it will incur" (security prospectus, April 7, 1981, p. 49).

In the Haddam Neck case before FERC (see note 5), the group of unaffiliated utilities that bought the nuclear power wholesale and that testified together as a Customer Resale Group, characterized the nuclear company's estimates as having "gross uncertainty." The FERC staff agreed with the uncertainty but felt that some provision had to be made for an independent decommissioning fund not under the control of the company. FERC staff wanted to hedge against the possibility of the company's walking away from the expired plant, leaving decommissioning to the public. The hearing judge, in his decision, stated that the data were bad and the methodology nonexistent but that some financial provision had to be made for decommissioning. The NRC, he wrote, had the responsibility of specifying decommissioning methods and financing but has not done so. The prospects for a repository, he stated, were bleak.

The 1977 congressional hearings on nuclear-power costs provided a forum for advocates and opponents on the committee to question expert witnesses.[17] The NRC testified that the technology "exists to safely decommission a nuclear reactor. . . . Any technical problems can be overcome." To this, the chairman of the committee, Mr. Ryan, stated that he "had never heard anyone in the Federal government who didn't tell him that 'everything was just great.' " The committee's report concluded that after thirty years, the technology to dismantle a large commercial reactor "has not yet been demonstrated, and the costs . . . are still unknown" (pp. 24 and 74).

Congressman Fountain, who is pro-nuclear, stated that "a myriad of questions remains unanswered, but some solutions are being worked on" (p. 78). A group of fourteen pro-nuclear congressmen, in a common dissent, agreed that "the report is correct in criticising 30 years of neglect" (p. 96).

The head of the General Accounting Office's energy staff, Monte Canfield, Jr., testified that geological formations for waste repositories were uncertain and that no technology exists to handle high-level wastes: "Little real progress has been made in solving the high-level waste storage problem." The time schedules of repositories were uncertain. "The [NRC's] idea that we know enough right now to just store it permanently is a little frightening" (pp. 13–17).

The National Resources Defense Council proposed that no further licensing of nuclear power plants be allowed because understanding of waste disposal is inadequate and multiplication of wastes is too great to be controlled: "A definitive finding of safety, as required by law, cannot be made at this time." It referred to "NRC's extraordinary oversimplification of the disposal problem," that radwaste will be safely stored because it is a necessity. The "same engineers" of the past thirty years, it testified, were repeating "engineering arrogance." The NRC, it said, was recalcitrant, and a moratorium on licensing should be called until safety of disposal is assured (pp. 131 ff.).

The commissioner of the California Energy Resources, Conservation and Development Commission, Emilio Vaanini III, testified: "The cardinal observation . . . which provides the essential perspective for all subsequent judgments, is that the expertise and information about what is going on is closely held—essentially by a cadre of 14 or 15 people in the energy establishment, and embodied in ERDA/NRC. . . . Worse, the information held by the technicians is subject to manipulation . . . by management to achieve preordained policy goals. . . . [There is a] legacy of mismanagement and lack of honesty. . . . Waste disposal . . . has to be depoliticized" (pp. 152–158).

Notes

1. U.S. Atomic Energy Commission, *The Nuclear Industry 1974,* WASH 1174-74 (Washington, D.C.: Government Printing Office, 1974), p. 64.

2. An excellent overview of waste disposal and the uncertainties is found in Fred C. Shapiro's *Radwaste* (New York: Random House, 1981), which was first written for *New Yorker* magazine.

3. Definitions have been standardized and will be found, for example, in the Nuclear Regulatory Commission's *Draft Generic Environmental Impact Statement on Decommissioning of Nuclear Facilities* NUREG-0586, Washington, D.C., January 1981, glossary (app. G).

4. Ibid., p. 2-6, and p. 0–10 ff.

5. Federal Energy Regulatory Commission, Connecticut Yankee Power Company, Docket ER78-360, initial decision, January 23, 1980, p. 24.

6. U.S. Nuclear Regulatory Commission, *Draft Generic Environmental Impact Statement,* NUREG-0586, p. 4–2.

7. Conversations with staff. U.S. Department of Energy/Energy Information Administration.

8. U.S. Nuclear Regulatory Commission, *Draft Generic Environmental Impact Statement,* NUREG-0586, foreword.

9. Ibid., p. 0–45.

10. Federal Energy Regulatory Commission, Connecticut Yankee Power Company.

11. U.S. Department of Energy/Energy Information Administration, *Nuclear Power Regulation, Energy Policy Study,* vol. 10 (Washington, D.C.: Government Printing Office, May 1980), p. 171.

12. Price indexes from Federal Reserve Board, *Bulletin,* March 1981, p. A44.

13. Nuclear Information and Resource Service, *Groundswell,* January/February 1982, p. 6.

14. By phone from Nuclear Information and Resource Service, Washington, D.C.

15. *Nuclear Power Costs,* Hearings before U.S. House of Representatives, Committee on Government Operations, September 1977, pp. 808, 812, and 1654; and followup letter of March 1, 1978.

16. *Nuclear Power Costs,* part 1, p. 585.

17. *Nuclear Power Costs.*

Part III
Case-Study Results

10 Introduction to Part III

Overview

Part II described the general analytical approach (including a description of the four case studies) and the rationale and methodology for the specific adjustments to the case studies. The following chapters outline for each case study how the adjustments were calculated and the resulting costs.

The following section summarizes the combined results of the four case-study analyses. Unless otherwise indicated, tables and numbers refer to the base case (see table 5–1 for case definitions). It should also be remembered that no adjustments are included for replacement power, fuel prices, major repairs, the Yoyo Factor, senescence, waste disposal, and decommissioning.

Summary of Results

The original and adjusted busbar costs (in mills/kwh) are summarized in table 10–1 for lifetime costs and in table 10–2 for initial commerciality costs (the AEC and Exxon case studies only). The major point to be gleaned from tables 10–1 and 10–2 is the diversity of costs among the four case studies, much of which is due to differences in time frame (for example, when the study was conducted and assumed commerciality dates) and methodology. Therefore, the absolute values of the costs are not comparable across case studies.

The ratios of coal to nuclear costs (table 10–3) provide the meat of the analysis. Before adjustments, in all three government studies (that is, AEC, ERDA, and NRC), coal's total average lifetime costs exceed nuclear's— coal ranges from 15 percent to 60 percent more expensive. Exxon's original numbers indicate that nuclear and high-sulfur coal with FGD cost approximately the same; low-sulfur coal without FGD is, however, 16 percent cheaper than nuclear. At initial commerciality, as opposed to lifetime, nuclear's total-cost advantage is considerably smaller in the AEC case study and disappears in the Exxon high-sulfur-coal scenario.

Table 10–1
Four Case Studies: Original and Adjusted Lifetime Busbar Costs
(mills/kwh)

Cost Category	Case Study	Original[a] Nuclear	Original[a] Coal	Adjusted[b] Nuclear	Adjusted[b] Coal
Capital	AEC	15.50	13.00	36.60	12.32
	ERDA	8.03	7.03	15.88	5.65
	NRC	45.50	38.00	95.90	29.43
	Exxon Hi-S coal with FGD	40.30	26.50	75.52	22.80
	Exxon Low-S coal without FGD	40.30	18.40	75.52	16.79
Fuel	AEC	14.70	31.90	14.70	27.95
	ERDA	7.79	11.53	7.79	10.88
	NRC	24.36	39.38	24.36	36.23
	Exxon Hi-S coal with FGD	22.00	33.80	22.00	33.01
	Exxon Low-S coal without FGD	22.00	32.60	22.00	32.60
O&M	AEC	3.90	9.70	12.70	9.70
	ERDA	1.35	2.69	3.33	2.69
	NRC	6.23	10.13	13.74	10.13
	Exxon Hi-S coal with FGD	4.00	7.00	9.50	7.00
	Exxon Low-S coal without FGD	4.00	4.60	9.50	4.60
Total	AEC	34.10	54.60	64.00	49.97
	ERDA	17.17	21.25	27.00	19.22
	NRC	76.29	87.51	134.00	75.79
	Exxon Hi-S coal with FGD	66.30	67.30	107.02	62.81
	Exxon Low-S coal without FGD	66.30	55.60	107.02	53.99

[a]The original AEC and Exxon data have been converted from costs at initial commerciality to lifetime costs as explained in the text.

[b]Adjustments are not included for replacement power, fuel prices, major repairs, the Yoyo Factor, senescence, waste disposal, and decommissioning.

After the adjustments are applied to total average costs, coal is cheaper than nuclear in all four case studies. High-sulfur coal with FGD ranges from 43 to 22 percent less expensive than nuclear. Low-sulfur coal without FGD is half as expensive as nuclear. Coal's cost advantage is greater at initial commerciality than over unit lifetime.

Looking at the individual cost components, coal's capital costs are lower than nuclear's both before and after adjustments. The original capital-cost ratios in the three government studies are very similar—coal is approximately 15 percent cheaper than nuclear. High-sulfur coal's capital-cost advantage is greater in the original Exxon numbers—34 percent—and low-sulfur coal is 54 percent cheaper than nuclear. After adjustments, the capital-cost ratios for the four high-sulfur coal scenarios are remarkably close—high-sulfur coal is approximately 33 percent cheaper than nuclear. Low-sulfur coal's capital costs are 78 percent lower than nuclear's.

Table 10–2
AEC and Exxon Case Studies: Original and Adjusted Total Average
Busbar Costs during Initial Commercial Operation
(mills/kwh)

	Original		Adjusted[a]	
	Nuclear	*Coal*	*Nuclear*	*Coal*
AEC[b]	22.60	28.90	47.05	26.71
Exxon Hi-S coal with FGD[c]	52.70	49.80	91.04	45.65
Exxon Low-S coal with FGD[c]	52.70	39.60	91.04	37.99

[a]Adjustments are not included for replacement power, fuel prices, major repairs, the Yoyo Factor, senescence, waste disposal, and decommissioning.

[b]AEC's costs are at initial day of commerciality.

[c]Exxon's costs are for first full year of commerciality.

Table 10–3
Four Case Studies: Original and Adjusted Average Lifetime Busbar
Costs: Coal as a Percentage of Nuclear

Cost Category	Case Study	Original[a]	Adjusted[b]
Capital	AEC	83.9%	33.7%
	ERDA	87.5	35.6
	NRC	83.5	30.7
	Exxon Hi-S coal with FGD	65.8	30.2
	Exxon Low-S coal without FGD	45.7	22.2
Fuel	AEC	217.0	190.1
	ERDA	148.0	139.7
	NRC	161.7	148.7
	Exxon Hi-S coal with FGD	153.6	150.0
	Exxon Low-S coal without FGD	148.2	148.2
O&M	AEC	248.7	76.4
	ERDA	199.3	80.8
	NRC	162.6	73.7
	Exxon Hi-S coal with FGD	175.0	73.7
	Exxon Low-S coal without FGD	115.0	48.4
Total[c]	AEC	160.1 (127.9)	78.1 (56.8)
	ERDA	123.8	71.2
	NRC	114.7	56.6
	Exxon Hi-S coal with FGD	101.5 (94.5)	58.7 (50.1)
	Exxon Low-S coal without FGD	83.9 (75.1)	50.4 (41.7)

[a]The original AEC and Exxon data have been converted from costs at initial commerciality to lifetime costs as explained in the text.

[b]Adjustments are not included for replacement power, fuel prices, major repairs, the Yoyo Factor, senescence, waste disposal, and decommissioning.

[c]Numbers in parentheses are for initial day of commerciality in the AEC case study and first year of commerciality in the Exxon case study.

Table 10–4
Four Case Studies:
Summary of Sensitivity Analysis

Capacity Factor (%)		Total Average Lifetime Costs: Coal as Percentage of Nuclear				
Nuclear	Coal	AEC	ERDA	NRC	Exxon Hi-S	Exxon Lo-S
55	70	78.1	71.2	56.6	58.7	50.4
60	70	83.4	75.7	60.7	62.8	54.0
50	70	72.5	66.5	52.3	54.4	46.7
55	75	76.8	69.8	55.1	57.3	49.4
55	65	79.6	72.8	58.2	60.3	51.7

Lifetime fuel costs for coal are higher than for nuclear in the original and adjusted case studies. Before adjustments, coal's fuel costs exceed nuclear's by 48 percent to 117 percent. After adjustments, the range narrows—coal is 40 percent to 90 percent more expensive than nuclear.

Consistent with the conventional but incorrect belief concerning relative O&M costs, the original numbers in all four case studies indicate that nuclear is less expensive than coal. O&M for high-sulfur coal ranges from 63 percent to 149 percent more expensive than for nuclear. Low-sulfur coal's O&M is 15 percent higher than nuclear's. After adjustments, the relationship reverses and the range narrows across studies. High-sulfur coal runs from 26 to 19 percent cheaper than nuclear; low-sulfur coal is 52 percent cheaper than nuclear.

The sensitivity analysis, summarized in table 10–4, examines the effects of raising and lowering the capacity-factor assumptions. None of the capacity-factor changes affected the overall results substantially. The largest impact results from different nuclear capacity factors. Increasing nuclear's capacity factor to 60 from 55 percent decreases coal's overall cost advantage by approximately 7 percent; lowering nuclear's capacity factor to 50 percent increases coal's cost advantage by approximately 7 percent. Changing coal's capacity factor affects the overall cost relationship between coal and nuclear very little. A higher capacity factor (75 versus 70 percent) for coal increases its cost advantage approximately 2 percent, and a lower capacity factor (65 percent) decreases its cost advantage by approximately 3 percent.

11 The AEC Case Study

The AEC, in *The Nuclear Industry 1974,* developed costs for nuclear and coal units as of the moment of commercialization in 1982. The first task then was to derive subsequent lifetime costs. The adjustments outlined in part II were applied to the initial commerciality and lifetime costs. The results are summarized here followed by a detailed description of the calculations. It should be remembered that adjustments are not included for replacement power, fuel prices, major repairs, the Yoyo Factor, senescence, waste disposal, and decommissioning. Case-study background, the overall analytical approach, and the rationale for each adjustment were discussed in part II.

Summary of Results

The original and adjusted initial commerciality costs are presented in table 11-1. Before adjustments, coal power is 28 percent more expensive than nuclear power; after adjustments, coal is 43 percent cheaper than nuclear. The original and adjusted capital costs for coal are lower than for nuclear, but the differential widens considerably after adjustments (16 to 66 percent cheaper). Conversely, coal's fuel costs are higher before and after adjustments, but the differential narrows after adjustments (118 to 91 percent more expensive). The relationship between nuclear and coal O&M costs reverses after adjustments; coal moves from being 147 percent more expensive than nuclear to being 24 percent cheaper.

The lifetime costs were developed using the AEC's own assumptions. Capital costs are constant over the life of a unit whereas fuel and O&M costs vary. Fuel and O&M unadjusted commerciality costs were escalated at AEC's 8 percent per year through 12.5 years which, being the midpoint of a nuclear unit's lifetime, approximates average lifetime costs. While a coal unit's lifetime midpoint is fifteen years, all lifetime costs are taken for 12.5 years of commerciality to preserve comparability.

The original and adjusted lifetime costs, presented in table 11-2, follow the same patterns as initial commerciality costs. The coal to nuclear ratios for the individual cost components—capital, fuel, and O&M—are almost

Table 11-1
AEC Case Study: Summary of Original and Adjusted Busbar Costs at Initial Commerciality for Future Nuclear and Coal Power Units
(mills/kwh)

	Original			Adjusted		
	Nuclear	*Coal*	*Coal as Percentage of Nuclear*	*Nuclear*	*Coal*	*Coal as Percentage of Nuclear*
Capital	15.50	13.00	83.9	36.60	12.32	33.7
Fuel	5.60	12.20	217.9	5.60	10.69	190.9
O&M	1.50	3.70	246.5	4.85	3.70	76.3
Total	22.60	28.90	127.9	47.05	26.71	56.8
CF	75%	75%		55%	70%	

Table 11-2
AEC Case Study: Summary of Original and Adjusted Lifetime Busbar Costs for Future Nuclear and Coal Power Units
(mills/kwh)

	Original			Adjusted		
	Nuclear	*Coal*	*Coal as Percentage of Nuclear*	*Nuclear*	*Coal*	*Coal as Percentage of Nuclear*
Capital	15.50	13.00	83.9	36.60	12.32	33.7
Fuel	14.70	31.90	217.0	14.70	27.95	190.1
O&M	3.90	9.70	248.7	12.70	9.70	76.4
Total	34.10	54.60	160.1	64.00	49.97	78.1
CF	75%	75%		55%	70%	

identical to the initial commerciality ratios. On total adjusted average busbar costs, coal's cost advantage decreases from 43 percent cheaper than nuclear at initial commerciality to 22 percent cheaper over the unit's life.

The sensitivity analysis for total average lifetime costs is summarized in table 11-3. The most significant effect on the coal-nuclear cost relationship results from nuclear capacity-factor changes. For the base case of a 55-percent capacity factor for nuclear and 70 percent for coal, coal is 22

Table 11–3
The AEC Case Study: Summary of Sensitivity Analysis

Capacity Factor (%)		Total Mills/kwh: Coal as Percentage of Nuclear
Nuclear	Coal	
55	70	78.1
60	70	83.4
50	70	72.5
55	75	76.8
55	65	79.6

percent cheaper than nuclear. The lower and higher nuclear capacity factors increase and decrease, respectively, coal's cost advantage by approximately 7 percent over the base case. Coal is 27 percent cheaper than nuclear in the $N = 50$ percent case and 17 percent cheaper in the $N = 60$ percent case. Changing coal's capacity factor affects the coal-nuclear cost ratios very little—approximately 2 percent above or below the base case. Coal is 23 percent cheaper than nuclear with a 75-percent capacity factor and 20 percent cheaper with a 65-percent capacity factor.

The calculations for the lifetime and initial commerciality costs just summarized are detailed in the sections that follow. Lifetime costs, the focus of the analysis, are addressed first, followed by the sensitivity analysis for lifetime costs. The adjustments to initial commerciality costs are then presented.

Adjustments to Lifetime Costs

(See worksheet in table 11–4.)

1. Technological Sufflation

   a. *Nuclear*

The 30-percent increase in capital costs is calculated as follows:

(Original capital)(0.30) = capital increase

(15.50)(0.30) = 4.65 mills/kwh

15.50 + 4.65 = 20.15 mills/kwh = total adjusted capital

Table 11–4
AEC Case Study: Worksheet for Adjusted Lifetime Busbar Costs of Future Nuclear and Coal Power Units
(mills/kwh)

		Nuclear	Coal
A.	Original Costs Adjusted for Lifetime		
	Capital	15.50	13.00
	Fuel	14.70	31.90
	O&M	3.90	9.70
	Total	34.10	54.60
	Capacity factor	75%	75%
B.	Capital Adjustments	55% CF	70% CF
	1. Technological sufflation	+4.65	
	Adjusted total, capital	20.15	
	2. Construction time	+2.22	−1.50
	Adjusted total, capital	22.37	11.50
	3. Unit life	+4.47	
	Adjusted total, capital	26.84	
	4. Capacity factor	+9.76	+0.82
	Adjusted total, capital	36.60	12.32
C.	Fuel Adjustments		
	5. Btu/lb of coal		−2.93
	6. Heat rate		−1.02
	Adjusted total, fuel	14.70	27.95
D.	O&M Adjustments		
	7. O&M	+8.80	
	Adjusted total, O&M	12.70	9.70
E.	Other Adjustments		
	8. Yoyo Factor		
	9. Replacement power		
	10. Major repairs		
	11. Senescence		
	12. Fuel prices		
	13. Use of low-sulfur coal		
	14. Waste disposal		
	15. Decommissioning		
F.	Sum of Adjustments[a]	+29.90	−4.63
G.	Adjusted Total Cost[a]	64.00	49.97
H.	Coal as Percent of Nuclear		78.08

[a]Adjustments are not included for items 8 through 15.

b. *Coal*
   No adjustment.

2. *Construction Time*

   a. *Nuclear*
      Raising nuclear's construction time from eight to ten years increases interest during construction (IDC) and pure price-level escalation during construction (EDC) by approximately 25 percent (that is, 2/8). The adjustment is first calculated in dollars per kw and then applied to the mills per kwh number as follows:

      (IDC + EDC in $/kw)(0.25) = $/kw capital increase

      ($300)(0.25) = $75.00/kw

      ($/kw capital increase)/($/kw total capital) = % increase in total capital

      $75/$680 = 0.1103

      (% increase in total capital)(mills/kwh total capital) = mills/kwh capital increase

      (0.1103)(20.15) = 2.22 mills/kwh

      20.15 + 2.22 = 22.37 mills/kwh = total adjusted capital

   b. *Coal*
      Reducing coal's construction time from 6.5 to 4.5 years decreases IDC and EDC by approximately 30.77 percent (that is, 2/6.5). The adjustment is first calculated in dollars per kw and then applied to the mills per kwh number as follows:

      (IDC + EDC in $/kw)(0.3077) = $/kw capital decrease

      ($213.5)(0.3077) = $147.81/kw

      ($/kw capital decrease)/($/kw total capital) = % decrease in total capital

      $65.69/$570 = 0.1152

      (% decrease in total capital)(mills/kwh total capital) = mills/kwh capital decrease

      (0.1152)(13.00) = 1.50 mills/kwh

13.00 − 1.50 = 11.50 mills/kwh = total adjusted capital

3. *Unit Life*

   a. *Nuclear*
   Reducing nuclear's unit life from thirty to twenty-five years increases the capital costs by 20 percent (that is 5/25) as follows:

   (Total capital)(0.20) = capital increase

   (22.37)(0.20) = 4.47 mills/kwh

   22.37 + 4.47 = 26.84 mills/kwh = total adjusted capital

   b. *Coal.*
   No adjustment.

4. *Capacity Factor*

   a. *Nuclear*
   Lowering nuclear's capacity factor from 75 to 55 percent increases capital costs by 36.36 percent (that is, 20/55) as follows:

   (Total capital)(0.3636) = capital increase

   (26.84)(0.3636) = 9.76 mills/kwh

   26.84 + 9.76 = 36.60 mills/kwh = total adjusted capital

   b. *Coal*
   Reducing coal's capacity factor from 75 to 70 percent increases capital costs by 7.14 percent (that is, 5/70) as follows:

   (Total capital)(0.0714) = capital increase

   (11.50)(0.0714) = 0.82 mills/kwh

   11.50 + 0.82 = 12.32 mills/kwh = total adjusted capital

5. *Btu per Pound of Coal*

   a. *Nuclear*
   No adjustment.

b. *Coal*
The AEC assumed 10,900 Btu/lb of bituminous coal which is, however, an average for all coal, including subbituminous and lignite, used by utilities. The adjustment to 12,000 Btu/lb reduces coal-fuel costs by 9.17 percent (that is, 1,100/12,000) as follows:

(Original fuel)(0.0917) = fuel decrease

(31.90)(0.0917) = 2.93 mills/kwh

6. *Heat Rate*

a. *Nuclear*
No adjustment.

b. *Coal*
Reducing the AEC's assumed heat rate of 9,500 Btu/kwh to 9,200 decreases coal's fuel costs by 3.20 percent (that is, 300/9,500) as follows:

(Original fuel)(0.0320) = fuel decrease

(31.90)(0.0320) = 1.02 mills/kwh

(Original fuel) − (Btu/lb coal adjustment) − (heat rate adjustment) = total adjusted fuel

31.90 − 2.93 − 1.02 = 27.95 mills/kwh

7. *O&M*

a. *Nuclear*
The 1978 average O&M for large nuclear units (3.85 mills/kwh with an experienced average capacity factor of 55 percent) was escalated through 12.5 years at the AEC's annual rate of 8 percent. The result is a lifetime O&M of 12.70 mills/kwh or an 8.80-mill increase over the original cost of 3.9 mills/kwh.

b. *Coal*
No adjustment.

Sensitivity Analysis for Lifetime Costs

Nuclear Capacity Factor

Changing nuclear's capacity factor affects capital and O&M costs by increasing or reducing the number of kwh over which the costs are averaged. Table 11–5 presents the sensitivity results for the 60 and 50 percent capacity factors. The calculations are outlined here.

Capital Costs. Adjustment 4 on the worksheet in table 11–4 captures the capacity-factor change for capital costs. A 60-percent capacity factor increases capital costs by 25 percent over the AEC's original 75-percent capacity factor (that is, 15/60) as follows:

$$(\text{Total capital})(0.25) = \text{capital increase}$$

$$(26.84)(0.25) = 6.71 \text{ mills/kwh}$$

$$26.84 + 6.71 = 33.55 \text{ mills/kwh} = \text{total adjusted capital}$$

A 50-percent capacity factor increases capital costs by 50 percent over the original 75-percent capacity factor (that is 25/50) as follows:

$$(\text{Total capital})(0.50) = \text{capital increase}$$

$$(26.84)(0.50) = 13.42 \text{ mills/kwh}$$

$$26.84 + 13.42 = 40.26 \text{ mills/kwh} = \text{total adjusted capital}$$

It should be noted that the sensitivity analysis increases capital costs over the original 75-percent capacity factor. However, relative to the 55-percent base case, capital costs decrease in the $N = 60$ percent case and increase in the $N = 50$ percent case.

O&M Costs. The capacity-factor adjustment for O&M costs is made at item 7 on the table 11–4 worksheet. The 60-percent capacity factor decreases the base-case O&M but increases the AEC's original O&M as follows:

$$(55\% \text{ CF O\&M})(55/60) = \text{total adjusted O\&M}$$

$$(12.70)(0.9167) = 11.64 \text{ mills/kwh}$$

$$(\text{Total adjusted O\&M}) - (\text{original O\&M}) = \text{O\&M increase}$$

$$11.64 - 3.90 = 7.74 \text{ mills/kwh}$$

Table 11-5
AEC Case Study: Sensitivity Analysis on Nuclear Capacity Factor

	Nuclear Costs (mills/kwh)			Coal as Percentage of Nuclear[a]		
	N = 55%	N = 60%	N = 50%	N = 55%	N = 60%	N = 50%
Capital	36.60	33.55	40.26	33.7	36.7	30.6
Fuel	14.70	14.70	14.70	190.1	190.1	190.1
O&M	12.70	11.64	13.97	76.4	83.3	69.4
Total	64.00	59.89	68.93	78.1	83.4	72.5

[a]Assumes a 70-percent capacity factor for coal.

The 50-percent capacity factor increases the base case and the AEC's original O&M as follows:

$$(55\% \text{ CF O\&M})(55/50) = \text{ total adjusted O\&M}$$

$$(12.70)(1.10) = 13.97 \text{ mills/kwh}$$

$$(\text{Total adjusted O\&M}) - (\text{original O\&M}) = \text{O\&M increase}$$

$$13.97 - 3.90 = 10.07 \text{ mills/kwh}$$

Coal Capacity Factor

Changing coal's capacity factor affects capital costs only by reducing the number of kwh over which costs are averaged (item 4 on the table 11-4 worksheet). The results are presented in table 11-6 and explained as follows:

Capital Costs. Increasing coal's capacity factor to 75 percent decreases capital costs relative to the 70-percent base case but requires no adjustment to the original AEC assumption of 75 percent. A 65-percent capacity factor increases capital costs by 15.38 percent over the original AEC costs (that is, 10/65) as follows:

$$(\text{Total capital})(0.1538) = \text{capital increase}$$

$$(11.50)(0.1538) = 1.77 \text{ mills/kwh}$$

$$11.50 + 1.77 = 13.27 \text{ mills/kwh} = \text{total adjusted capital}$$

Table 11-6
AEC Case Study: Sensitivity Analysis on Coal Capacity Factor

	Coal Costs (mills/kwh)			Coal as Percentage of Nuclear[a]		
	C = 70%	C = 75%	C = 65%	C = 70%	C = 75%	C = 65%
Capital	12.32	11.50	13.27	33.7	31.4	36.3
Fuel	27.95	27.95	27.95	190.1	190.1	190.1
O&M	9.70	9.70	9.70	76.4	76.4	76.4
Total	49.97	49.15	50.92	78.1	76.8	79.6

[a]Assumes a 55-percent capacity factor for nuclear.

Adjustments to Initial Commerciality Costs

The adjustments outlined in part II were applied to the initial commerciality costs as well as to the lifetime costs. The capital-cost adjustments are identical for initial commerciality and lifetime and are not therefore outlined again here. The fuel and O&M adjustments are conceptually the same for lifetime and initial commerciality but are applied on a lower base for the latter. The worksheet for the adjusted commerciality costs is contained in table 11-7, and the fuel and O&M adjustments are explained as follows.

1. through 4.   Capital Adjustments

Same as for lifetime costs.

5.   Btu per Pound of Coal

a.   *Nuclear*
No adjustment.

b.   *Coal*
The adjustment from 10,900 Btu/lb of coal to 12,000 Btu (explained under lifetime costs) reduces coal fuel costs by 9.17 percent (that is, 1,100/12,000) as follows:

(Original fuel)(0.0917) = fuel decrease

(12.20)(0.0917) = 1.12 mills/kwh

12.20 − 1.12 = 11.08 mills/kwh

Table 11-7
AEC Case Study: Worksheet for Adjusted Busbar Costs at Initial
Commerciality of Future Nuclear and Coal Power Units
(mills/kwh)

		Nuclear	*Coal*
A.	Original Costs		
	Capital	15.50	13.00
	Fuel	5.60	12.20
	O&M	1.50	3.70
	Total	22.60	28.90
	Capacity factor	75%	75%
B.	Capital Adjustments	55% CF	70% CF
	1. Technological sufflation	+4.65	
	Adjusted total, capital	20.15	
	2. Construction time	+2.22	−1.50
	Adjusted total, capital	22.37	11.50
	3. Unit life	+4.47	
	Adjusted total, capital	26.84	
	4. Capacity factor	+9.76	+0.82
	Adjusted total, capital	36.60	12.32
C.	Fuel Adjustments		
	5. Btu/lb of coal		−1.12
	6. Heat rate		−0.39
	Adjusted total, fuel	5.60	10.69
D.	O&M Adjustments		
	7. O&M	+3.35	
	Adjusted total, O&M	4.85	3.70
E.	Other Adjustments		
	8. Yoyo Factor		
	9. Replacement power		
	10. Major repairs		
	11. Senescence		
	12. Fuel prices		
	13. Use of low-sulfur coal		
	14. Waste disposal		
	15. Decommissioning		
F.	Sum of Adjustments[a]	+24.45	−2.19
G.	Adjusted Total Cost[a]	47.05	26.71
H.	Coal as Percent of Nuclear		56.77

[a]Adjustments are not included for items 8 through 15.

6. *Heat Rate*

    a. *Nuclear*
        No adjustment.

    b. *Coal*
        The adjustment from 9,500 to 9,200 Btu/kwh decreases coal's fuel costs by 3.16 percent (that is, 300/9,500) as follows:

    (Original fuel)(0.0316) = fuel decrease

    (12.20)(0.0316) = 0.39 mills/kwh

    (Original fuel) − (Btu/lb coal adjustment) − (heat rate adjustment) = total adjusted fuel

    12.20 − 1.12 − 0.39 = 10.69 mills/kwh

7. *O&M*

    a. *Nuclear*
        The 1978 average O&M for large nuclear units (3.85 mills/kwh) was escalated to 1982 (commerciality) at the AEC's annual rate of 8 percent. The result is a commerciality O&M of 4.85 mills/kwh, or a 3.35-mill increase over the original cost of 1.5 mills/kwh.

    b. *Coal*
        No adjustment.

12 The ERDA Case Study

The ERDA case study is based on the report titled "Comparing New Technologies for the Electric Utilities," December 1976, and the generous assistance (via phone and letter) of the study's project leader, Les Goudarzi. The great difficulties with the published study that necessitated Goudarzi's input are discussed in part II. The ERDA study refers, therefore, to the reconstructed report, that is, the 1976 study as elaborated and clarified by Goudarzi. The study provides levelized, present value, average life-cycle costs (1982 commerciality) for the Southeastern Region (SERC), which Goudarzi deemed most representative of costs nationwide.

The original and adjusted costs, including sensitivity analysis, are summarized here followed by a detailed account of the calculations. It should be remembered that adjustments are not included for replacement power, fuel prices, major repairs, the Yoyo Factor, senescence, waste disposal, and decommissioning. Case-study background, the overall analytical approach, and the rationale for each adjustment are discussed in part II.

Summary of Results

The original and adjusted average lifetime costs are summarized in table 12-1. Before adjustments, coal power is 24 percent more expensive than nuclear power; after adjustments, coal is 29 percent cheaper. The original and adjusted capital costs for coal are lower than for nuclear, but the differential widens considerably after adjustments (from 12 to 64 percent cheaper). Conversely, coal's fuel costs are higher than nuclear's before and after adjustments, but the differential narrows (from 48 to 40 percent more expensive). The relationship between nuclear and coal O&M costs reverses after adjustments; coal moves from being 99 percent more expensive than nuclear to being 19 percent cheaper.

The sensitivity analysis is summarized in table 12-2. The most significant impact on the coal/nuclear cost relationship results from nuclear capacity-factor changes. In the base case of a 55-percent capacity factor for nuclear and 70 percent for coal, coal is 29 percent cheaper than nuclear. The 50-percent nuclear capacity factor increases coal's cost advantage by 7 percent over the base case—coal is 33 percent cheaper than nuclear. The 60-per-

119

Table 12-1
ERDA Case Study: Summary of Original and Adjusted Lifetime Busbar Costs for Future Nuclear and Coal Power Units
(mills/kwh)

	Original			*Adjusted*		
	Nuclear	*Coal*	*Coal as Percentage of Nuclear*	*Nuclear*	*Coal*	*Coal as Percentage of Nuclear*
Capital	8.03	7.03	87.5	15.88	5.65	35.6
Fuel	7.79	11.53	148.0	7.79	10.88	139.7
O&M	1.35	2.69	199.3	3.33	2.69	80.8
Total	17.17	21.25	123.8	27.00	19.22	71.2
CF	65%	58%		55%	70%	

Table 12-2
ERDA Case Study: Summary of Sensitivity Analysis

	Capacity Factor (%)	*Total Mills/kwh: Coal as Percentage of Nuclear*
Nuclear	*Coal*	
55	70	71.2
60	70	75.7
50	70	66.5
55	75	69.8
55	65	72.8

cent nuclear capacity factor decreases coal's cost advantage by 6 percent over the base case—coal is 24 percent cheaper than nuclear. The higher and lower coal capacity factors change the coal/nuclear cost ratios by only 2 percent in either direction; coal is 30 percent cheaper with a 75-percent capacity factor and 27 percent cheaper at 65 percent.

The calculations for the lifetime costs, including sensitivity analysis, are detailed in the following sections.

Adjustments to Lifetime Costs

(See worksheet in table 12-3.)

Table 12–3
ERDA Case Study: Worksheet for Adjusted Lifetime Busbar Costs of
Future Nuclear and Coal Power Units
(mills/kwh)

		Nuclear	*Coal*
A.	Original Costs		
	Capital	8.03	7.03
	Fuel	7.79	11.53
	O&M	1.35	2.69
	Total	17.17	21.25
	Capacity factor	65%	58%
B.	Capital Adjustments	55% CF	70% CF
	1. Technological sufflation	+ 2.41	
	Adjusted total, capital	10.44	
	2. Construction time	+ 0.76	− 0.21
	Adjusted total, capital	11.20	6.82
	3. Unit life	+ 2.24	
	Adjusted total, capital	13.44	
	4. Capacity factor	+ 2.44	− 1.17
	Adjusted total, capital	15.88	5.65
C.	Fuel Adjustments		
	5. Btu/lb of coal		
	6. Heat rate		− 0.65
	Adjusted total, fuel	7.79	10.88
D.	O&M Adjustments		
	7. O&M	+ 1.98	
	Adjusted total, O&M	3.33	2.69
E.	Other Adjustments		
	8. Yoyo Factor		
	9. Replacement power		
	10. Major repairs		
	11. Senescence		
	12. Fuel prices		
	13. Use of low-sulfur coal		
	14. Waste disposal		
	15. Decommissioning		
F.	Sum of Adjustments[a]	+ 9.83	− 2.03
G.	Adjusted Total Cost[a]	27.00	19.22
H.	Coal as Percent of Nuclear		71.19

[a]Adjustments are not included for items 8 through 15.

1.   Technological Sufflation

a. *Nuclear*
The 30-percent increase in capital costs is calculated as follows:

(Original capital)(0.30) = capital increase

(8.03)(0.30) = 2.41 mills/kwh

8.03 + 2.41 = 10.44 mills/kwh = total adjusted capital

b. *Coal*
No adjustment.

2.   Construction Time

a. *Nuclear*
Raising nuclear's construction time from 7 to 10.5 years increases interest during construction (IDC) by approximately 50 percent (that is, 3.5/7). The levelized, present-value method obviates the need for an inflation adjustment (EDC). The adjustment is first calculated in dollars per kw and then applied to the mills per kwh number as follows:

(IDC in $/kw)(0.50) = $/kw capital increase

($85.5)(0.50) = $42.75/kw

($/kw capital increase)/($/kw total capital) = % increase in total capital

$42.75/$588.5 = 0.0726

(% increase in total capital)(mills/kwh total capital) = mills/kwh capital increase

(0.0726)(10.44) = 0.76 mills/kwh

10.44 + 0.76 = 11.20 = total adjusted capital

b. *Coal*
Reducing coal's construction time from 6 to 4.5 years decreases IDC by approximately 25 percent (that is, 1.5/6). The adjustment is first calculated in dollars per kw and then applied to the mills per kwh number as follows:

(IDC in \$/kw)(0.25) = \$/kw capital decrease

(\$57.5)(0.25) = \$14.37/kw

(\$/kw capital decrease)/(\$/kw total capital) = % decrease in total capital

\$14.37/\$478.5 = 0.0300

(% decrease in total capital)(mills/kwh total capital) = mills/kwh capital decrease

(0.0300)(7.03) = 0.21 mills/kwh

7.03 − 0.21 = 6.82 mills/kwh = total adjusted capital

3. *Unit Life*

   a. *Nuclear*
      Reducing nuclear's unit life from thirty to twenty-five years increases the capital costs by 20 percent (that is, 5/25) as follows:

      (Total capital)(0.20) = capital increase

      (11.20)(0.20) = 2.24 mills/kwh

      11.20 + 2.24 = 13.44 mills/kwh = total adjusted capital

   b. *Coal*
      No adjustment.

4. *Capacity Factor*

   a. *Nuclear*
      Lowering nuclear's capacity factor from 65 to 55 percent increases capital costs by 18.18 percent (that is, 10/55) as follows:

      (Total capital)(0.1818) = capital increase

      (13.44)(0.1818) = 2.44 mills/kwh

      13.44 + 2.44 = 15.88 mills/kwh = total adjusted capital

   b. *Coal*
      Increasing coal's capacity factor to 70 from 58 percent decreases coal's capital costs by 17.14 percent (that is, 12/70) as follows:

(Total capital)(0.1714) = capital decrease

(6.82)(0.1714) = 1.17 mills/kwh

6.82 − 1.17 = 5.65 mills/kwh = total adjusted capital

5. *Btu per Pound of Coal*

   a. *Nuclear*
   No adjustment.

   b. *Coal*
   A Btu/lb of coal adjustment is not possible as fuel costs are expressed in cents per million Btu.

6. *Heat Rate*

   a. *Nuclear*
   No adjustment.

   b. *Coal*
   Reducing ERDA's assumed heat rate from 9,751 to 9,200 Btu/kwh decreases coal's fuel costs by 5.65 percent (that is, 551/9,751) as follows:

   (Original fuel)(0.0565) = fuel decrease

   (11.53)(0.0565) = 0.65 mills/kwh

   11.53 − 0.65 = 10.88 mills/kwh = total adjusted fuel

7. *O&M*

   a. *Nuclear*
   The 1978 average O&M for large nuclear units (3.85 mills/kwh with an average capacity factor of 55 percent) was discounted back to 1975 at ERDA's 5-percent escalation rate. The result is 3.33 mills/kwh, which represents a 1.98-mill increase over ERDA's original cost of 1.35 mills.

   b. *Coal*
   No adjustment.

Sensitivity Analysis for Lifetime Costs

Nuclear Capacity Factor

Changing nuclear's capacity factor affects capital and O&M costs by increasing or decreasing the number of kwh over which the costs are averaged. Table 12–4 presents the sensitivity results for the 60-percent and 50-percent capacity factors. The calculations are outlined as follows.

Capital Costs. Adjustment 4 on the worksheet in table 12–3 captures the capacity-factor change for capital costs. A 60-percent capacity factor increases capital costs by 8.33 percent over ERDA's assumed 65-percent capacity factor (that is, 5/60) as follows:

$$(\text{Total capital})(0.0833) = \text{capital increase}$$

$$(13.44)(0.0833) = 1.12 \text{ mills/kwh}$$

$$13.44 + 1.12 = 14.56 \text{ mills/kwh} = \text{total adjusted capital}$$

A 50-percent capacity factor increases capital costs by 30.00 percent over the 65-percent capacity-factor assumption (that is, 15/50) as follows:

$$(\text{Total capital})(0.3000) = \text{capital increase}$$

$$(13.44)(0.3000) = 4.03 \text{ mills/kwh}$$

$$13.44 + 4.03 = 17.47 \text{ mills/kwh} = \text{total adjusted capital}$$

It should be remembered that the sensitivity analysis increases capital costs over ERDA's original 65-percent capacity factor. However, relative to the base case, capital costs decrease in the $N = 60$ percent case and increase in the $N = 50$ percent case.

O&M Costs. The capacity-factor adjustment for O&M costs is made at item 7 on the table 12–3 worksheet. The 60-percent capacity factor decreases O&M over the base case but increases O&M over ERDA's original cost, as follows:

$$(55\% \text{ CF O\&M})(55/60) = \text{total adjusted O\&M}$$

$$(3.33)(0.9167) = 3.05 \text{ mills/kwh}$$

$$(\text{Total adjusted O\&M}) - (\text{original O\&M}) = \text{O\&M increase}$$

$$3.05 - 1.35 = 1.70 \text{ mills/kwh}$$

Table 12-4
ERDA Case Study: Sensitivity Analysis on Nuclear Capacity Factor

| | Nuclear Costs (mills/kwh) | | | Coal as Percentage of Nuclear[a] | | |
	N = 55%	N = 60%	N = 50%	N = 55%	N = 60%	N = 50%
Capital	15.88	14.56	17.47	35.6	38.8	32.3
Fuel	7.79	7.79	7.79	139.7	139.7	139.7
O&M	3.33	3.05	3.66	80.8	88.2	73.5
Total	27.00	25.40	28.92	71.2	75.7	66.5

[a]Assumes a 70-percent capacity factor for coal.

The 50-percent capacity factor increases O&M over the base case and over the ERDA's original O&M as follows:

(55% CF O&M)(55/50) = total adjusted O&M

(3.33)(1.10) = 3.66 mills/kwh

(Total adjusted O&M) − (original O&M) = O&M increase

3.66 − 1.35 = 2.31 mills/kwh

Coal Capacity Factor

Changing coal's capacity factor affects capital costs only by reducing the number of kwh over which costs are averaged (item 4 on the table 12-3 worksheet). The results are presented in table 12-5 and explained as follows:

Capital Costs. Increasing coal's capacity factor to 75 percent decreases capital costs by 22.67 percent over the original ERDA costs (17/75) as follows:

(Total capital)(0.2267) = capital decrease

(6.82)(0.2267) = 1.55 mills/kwh

6.82 − 1.55 = 5.27 mills/kwh = total adjusted capital

A 65-percent capacity factor decreases coal's capital costs by 10.77 percent over ERDA's estimate (7/65) as follows:

Table 12-5
ERDA Case Study: Sensitivity Analysis on Coal Capacity Factor

	Coal Costs (mills/kwh)			Coal as Percentage of Nuclear[a]		
	C = 70%	C = 75%	C = 65%	C = 70%	C = 75%	C = 65%
Capital	5.65	5.27	6.09	35.6	33.2	38.4
Fuel	10.88	10.88	10.88	139.7	139.7	139.7
O&M	2.69	2.69	2.69	80.8	80.8	80.8
Total	19.22	18.84	19.66	71.2	69.8	72.8

[a]Assumes a 55-percent capacity factor for nuclear.

(Total capital)(0.1077) = capital decrease

(6.82)(0.1077) = 0.73 mills/kwh

6.82 − 0.73 = 6.09 mills/kwh = total adjusted capital

The C = 75 percent case decreases capital costs over the base case whereas the C = 65 percent case increases costs over the base case.

13 The NRC Case Study

The NRC study of 1979, *Draft Environmental Statement, New England Power Units 1 and 2,* compared the economics of nuclear and coal generating units (1988 commerciality) for the proposed Charlestown, Rhode Island, plant site. The original and adjusted costs, including sensitivity analysis, are summarized here followed by a detailed account of the calculations. It should be remembered that adjustments are not included for replacement power, fuel prices, major repairs, the Yoyo Factor, senescence, waste disposal, and decommissioning. Case-study background, the overall analytical approach, and the rationale for each adjustment are discussed in part II.

Summary of Results

The original and adjusted average lifetime costs are summarized in table 13-1. Before adjustments, coal power is 15 percent more expensive than nuclear power; after adjustments, coal is 43 percent cheaper. The original and adjusted capital costs for coal are lower than for nuclear, but the differential widens considerably after adjustments (from 16 to 69 percent cheaper). Conversely, coal's fuel costs are higher than nuclear's before and after adjustments, but the differential narrows (from 62 to 49 percent more expensive). The relationship between nuclear and coal O&M costs reverses after adjustments; coal moves from being 63 percent more expensive than nuclear to being 26 percent cheaper.

The sensitivity analysis is summarized in table 13-2. The most significant impact on the coal/nuclear cost relationship results from nuclear capacity-factor changes. In the base case, a 55-percent capacity factor for nuclear and 70 percent for coal, coal is 43 percent cheaper than nuclear. The 50-percent nuclear capacity factor increases coal's cost advantage by 8 percent over the base case—coal is 48 percent cheaper than nuclear. The 60-percent nuclear capacity factor decreases coal's cost advantage by 7 percent over the base case—coal is 39 percent cheaper than nuclear. The higher and lower coal capacity factors change the coal/nuclear cost ratios by only 3 percent above or below the base case; coal is 45 percent cheaper with a

Table 13–1
NRC Case Study: Summary of Original and Adjusted Lifetime Busbar Costs for Future Nuclear and Coal Power Units
(mills/kwh)

	Original			*Adjusted*		
	Nuclear	*Coal*	*Coal as Percentage of Nuclear*	*Nuclear*	*Coal*	*Coal as Percentage of Nuclear*
Capital	45.50	38.00	83.5	95.90	29.43	30.7
Fuel	24.36	39.38	161.7	24.36	36.23	148.7
O&M	6.23	10.13	162.6	13.74	10.13	73.7
Total	76.29	87.51	114.7	134.00	75.79	56.6
CF	60%	60%		55%	70%	

Table 13–2
NRC Case Study: Summary of Sensitivity Analysis

Capacity Factor (%)		*Total Mills/kwh: Coal as Percentage of Nuclear*
Nuclear	*Coal*	
55	70	56.6
60	70	60.7
50	70	52.3
55	75	55.1
55	65	58.2

75-percent capacity factor and 42 percent cheaper with a 65-percent capacity factor.

The calculations for the lifetime costs, including sensitivity analysis, just summarized are detailed in the following sections.

Adjustments to Lifetime Costs

(See worksheet in table 13–3.)

1A. Unit-Size Adjustment

   a. *Nuclear*
      No adjustment.

Table 13–3
NRC Case Study: Worksheet for Adjusted Lifetime Busbar Costs of Future Nuclear and Coal Power Units
(mills/kwh)

		Nuclear	Coal
A.	Original Costs		
	Capital	45.50	38.00
	Fuel	24.36	39.38
	O&M	6.23	10.13
	Total	76.29	87.51
	Capacity factor	60%	60%
B.	Capital Adjustments	55% CF	70% CF
	1A. Unit size		− 3.66
	Adjusted total, capital		34.34
	1B. Technological sufflation	+ 13.71	
	Adjusted total, capital	59.41	
	2. Construction time	+ 13.85	
	Adjusted total, capital	73.26	
	3. Unit life	+ 14.65	
	Adjusted total, capital	87.91	
	4. Capacity factor	+ 7.99	− 4.91
	Adjusted total, capital	95.90	29.43
C.	Fuel Adjustments		
	5. Btu/lb of coal		
	6. Heat rate		− 3.15
	Adjusted total, fuel	24.36	36.23
D.	O&M Adjustments		
	7. O&M	+ 7.51	
	Adjusted total, O&M	13.74	10.13
E.	Other Adjustments		
	8. Yoyo Factor		
	9. Replacement power		
	10. Major repairs		
	11. Senescence		
	12. Fuel prices		
	13. Use of low-sulfur coal		
	14. Waste disposal		
	15. Decommissioning		
F.	Sum of Adjustments[a]	+ 57.71	− 11.72
G.	Adjusted Total Cost[a]	134.00	75.79
H.	Coal as Percent of Nuclear		56.56

[a]Adjustments are not included for items 8 through 15.

b. *Coal*

NRC compared the economics of two 1,150-MW nuclear units with three 767-MW coal units. The other three case studies compared nuclear and coal units of equal size. (Exxon presented a second analysis comparing two large nuclear units with three 800-MW coal units.) Given that scale problems are greater with large nuclear than with large coal units (see part I), it would seem more logical to compare large nuclear and large coal units in the NRC case as well.

The first task then was to convert NRC's capital costs from three 767-MW units to two 1,150-MW units. The Exxon case study because it had analyzed both large and small coal units, provided the analog for the conversion. The methodology for the conversion is described as follows:

Step 1: Derive ratio of NRC large coal unit costs to NRC large nuclear unit costs, in $/kw, using Exxon analog (after adjustments):

$$\left(\frac{\dfrac{\text{NRC small coal \$}}{\text{NRC large nuclear \$}}}{\dfrac{\text{Exxon small coal \$}}{\text{Exxon large nuclear \$}}} \right) \left(\frac{\text{Exxon large coal \$}}{\text{Exxon large nuclear \$}} \right)$$

$$= \frac{\text{NRC large coal \$}}{\text{NRC large nuclear \$}} \left(\frac{0.8316}{0.8000} \right)(0.7231) = 0.7516$$

Step 2: Using ratio derived in step 1, solve for NRC large coal unit cost in $/kw:

$$\frac{\text{NRC large coal \$}}{\text{NRC large nuclear \$}} = 0.7516$$

NRC large coal $ = (0.7516)(NRC large nuclear $)

$2076 = (0.7516)($2,762)

Step 3: Apply $/kw results to mills/kwh costs:

$$\left(\frac{\text{NRC large coal \$}}{\text{NRC small coal \$}} \right) (\text{NRC small coal mills/kwh} =$$

$$= (\text{total adjusted NRC large coal mills/kwh})$$

$$\left(\frac{\$2,076}{\$2,297} \right) (38.00 \text{ mills}) = 34.34 \text{ mills/kwh}$$

Step 4: Calculate increment to original NRC capital cost in mills/kwh:

(NRC small coal mills) − (NRC large coal mills) =

capital decrease

38.00 − 34.34 = 3.66 mills/kwh

1B. Technological Sufflation

a. *Nuclear*
   The 30-percent increase in capital costs is calculated as follows:

(Original capital)(0.30) = capital increase

(45.70)(0.30) = 13.71 mills/kwh

45.70 + 13.71 = 59.41 mills/kwh = total adjusted capital

b. *Coal*
   No adjustment.

2. *Construction Time*

a. *Nuclear*
   Raising nuclear's construction time from eight to twelve years
   increases interest during construction (IDC) and price-level escalation during construction (EDC) by approximately 50 percent (that
   is, 4/8). The adjustment is first calculated in dollars per kw and
   then applied to the mills per kwh number as follows:

(IDC + EDC in $/kw)(0.50) = $/kw capital increase

($1,288)(0.50) = $644.00/kw

($/kw capital increase)/($/kw total capital) =

% increase in total capital

$644/$2,762 = 0.2332

(% increase in total capital)(mills/kwh total capital) =

mills/kwh capital increase

(0.2332)(59.41) = 13.85 mills/kwh

59.41 + 13.85 = 73.26 mills/kwh = total adjusted capital

b.   *Coal*

No adjustment was necessary as the NRC assumed approximately 4.5 years for the construction of a coal unit.

3.   *Unit Life*

a.   *Nuclear*

Reducing nuclear's unit life from thirty to twenty-five years increases the capital costs by 20 percent (that is, 5/25) as follows:

(Total capital)(0.20) = capital increase

(73.26)(0.20) = 14.65 mills/kwh

73.26 + 14.65 = 87.91 mills/kwh = total adjusted capital

b.   *Coal*

No adjustment.

4.   *Capacity Factor*

a.   *Nuclear*

Lowering nuclear's capacity factor from 60 to 55 percent increases capital costs by 9.09 percent (that is, 5/55) as follows:

(Total capital)(0.0909) = capital increase

(87.91)(0.0909) = 7.99 mills/kwh

87.91 + 7.99 = 95.90 mills/kwh = total adjusted capital

b.   *Coal*

Increasing coal's capacity factor to 70 percent from 60 percent decreases coal's capital costs by 14.29 percent (that is, 10/70) as follows:

(Total capital)(0.1429) = capital decrease

(34.34)(0.1429) = 4.91 mills/kwh

34.34 − 4.91 = 29.43 mills/kwh = total adjusted capital

5. *Btu per Pound of Coal*

   a. *Nuclear*
      No adjustment.

   b. *Coal*
      No adjustment was made as the NRC assumed 12,000 Btu/lb of coal.

6. *Heat Rate*

   a. *Nuclear*
      No adjustment.

   b. *Coal*
      Reducing NRC's heat rate from 10,000 to 9,200 Btu/kwh decreases coal's fuel costs by 8.00 percent (that is, 800/10,000) as follows:

      (Original fuel)(0.0800) = fuel decrease

      (39.38)(0.0800) = 3.15 mills/kwh

      39.38 − 3.15 = 36.23 mills/kwh = total adjusted fuel

7. *O&M*

   a. *Nuclear*
      The complexity of NRC's methodology for computing O&M costs (a computer model that calculated discounted present-value lifetime costs) rendered the 1978 average O&M for large nuclear plants an inappropriate basis for adjusting the NRC number. The NRC number is clearly incorrect—while experienced coal O&M is in fact lower than nuclear O&M (see discussion in part II), the NRC shows coal to be 63 percent more expensive. The adjustment was made by accepting the NRC's O&M for coal (as in the other three case studies) and equating NRC's coal/nuclear ratio to Exxon's adjusted ratio in the high-sulfur-coal case. The adjustment was calculated as follows:

      $$\frac{\text{Exxon adjusted high-sulfur-coal O\&M}}{\text{Exxon adjusted nuclear O\&M}} = 0.737$$

$$\frac{\text{NRC adjusted coal O\&M}}{\text{NRC adjusted nuclear O\&M}} = 0.737$$

$$\frac{10.13}{X} = 0.737$$

$X = 13.74$ mills/kwh = total adjusted NRC nuclear O&M

(NRC adjusted O&M) $-$ (NRC original O&M) =

O&M increase

$13.74 = 6.23 = 7.51$ mills/kwh

b. *Coal*
No adjustment.

Sensitivity Analysis for Lifetime Costs

Nuclear Capacity Factor

Changing nuclear's capacity factor affects capital and O&M costs by increasing or reducing the number of kwh over which the costs are averaged. Table 13–4 presents the sensitivity results for the 60-percent and 50-percent capacity factors. The calculations are outlined as follows.

Capital Costs. Adjustment 4 on the worksheet in table 13–3 captures the capacity-factor change for capital costs. A 60-percent capacity factor does not require an adjustment as the NRC case originally assumed 60 percent.

Table 13–4
NRC Case Study: Sensitivity Analysis on Nuclear Capacity Factor

	Nuclear Costs (mills/kwh)			Coal as Percentage of Nuclear[a]		
	N = 55%	*N = 60%*	*N = 50%*	*N = 55%*	*N = 60%*	*N = 50%*
Capital	95.90	87.91	105.49	30.7	33.5	27.9
Fuel	24.36	24.36	24.36	148.7	148.7	148.7
O&M	13.74	12.60	15.11	73.7	80.4	67.0
Total	134.00	124.87	144.96	56.6	60.7	52.3

[a]Assumes a 70-percent capacity factor for coal.

A 50-percent capacity factor increases capital costs by 20 percent over the original case (that is, 10/50) as follows:

$$(\text{Total capital})(0.20) = \text{capital increase}$$

$$(87.91)(0.20) = 17.58 \text{ mills/kwh}$$

$$87.91 + 17.58 = 105.49 \text{ mills/kwh} = \text{total adjusted capital}$$

The 50-percent capacity factor also increases capital costs over the 55-percent base case.

O&M Costs. The capacity-factor adjustment for O&M costs is made at item 7 on the table 13-3 worksheet. The 60-percent capacity factor decreases the base case O&M but increases the NRC's original O&M as follows:

$$(55\% \text{ CF O\&M})(55/60) = \text{total adjusted O\&M}$$

$$(13.74)(0.9167) = 12.60 \text{ mills/kwh}$$

$$(\text{Total adjusted O\&M}) - (\text{original O\&M}) = \text{O\&M increase}$$

$$12.60 - 6.23 = 6.37 \text{ mills/kwh}$$

The 50-percent capacity factor increases O&M over the base case and the NRC's original O&M as follows:

$$(55\% \text{ CF O\&M})(55/50) = \text{total adjusted O\&M}$$

$$(13.74)(1.10) = 15.11 \text{ mills/kwh}$$

$$(\text{Total adjusted O\&M}) - (\text{original O\&M}) = \text{O\&M increase}$$

$$15.11 - 6.23 = 8.88 \text{ mills/kwh}$$

Coal Capacity Factor

Changing coal's capacity factor affects capital costs only by reducing the number of kwh over which costs are averaged (item 4 on the table 13-3 worksheet). The results are presented in table 13-5 and explained below.

Capital Costs. Increasing coal's capacity factor to 75 percent decreases capital costs by 20.00 percent over the original NRC assumption of 60 percent (that is 15/75) as follows:

Table 13-5
NRC Case Study: Sensitivity Analysis on Coal Capacity Factor

	Coal Costs (mills/kwh)			Coal as Percentage of Nuclear[a]		
	C = 70%	C = 75%	C = 65%	C = 70%	C = 75%	C = 65%
Capital	29.43	27.47	31.70	30.7	28.6	33.1
Fuel	36.23	36.23	36.23	148.7	148.7	148.7
O&M	10.13	10.13	10.13	73.7	73.7	73.7
Total	75.79	73.83	78.06	56.6	55.1	58.2

[a]Assumes a 55-percent capacity factor for nuclear.

$$(\text{Total capital})(0.2000) = \text{capital decrease}$$

$$(34.34)(0.2000) = 6.87 \text{ mills/kwh}$$

$$34.34 - 6.87 = 27.47 \text{ mills/kwh} = \text{total adjusted capital}$$

A 65-percent capacity factor increases capital costs by 7.69 percent over the original NRC cost (that is, 5/65) as follows:

$$(\text{Total capital})(0.0769) = \text{capital decrease}$$

$$(34.34)(0.0769) = 2.64 \text{ mills/kwh}$$

$$34.34 - 2.64 = 31.70 \text{ mills/kwh} = \text{total adjusted capital}$$

Relative to the base case, the $C = 65$ percent case increases, and the $C = 75$ percent case decreases capital costs.

14 The Exxon Case Study

The Exxon study of November 1977 (whose nuclear/coal ratios were reaffirmed for 1979) compared nuclear- and coal-power costs for the first year of commercial operation (1985) in New England. A unique contribution of the study is that cost projections were made for low-sulfur Appalachian coal without FGD as well as for high-sulfur coal with FGD. The first task was to derive lifetime costs. The adjustments outlined in part II were applied to the first year and lifetime costs. The results are summarized here followed by a detailed description of the calculations. It should be remembered that adjustments are not included for replacement power, fuel prices, major repairs, the Yoyo Factor, senescence, waste disposal, and decommissioning. Case-study background, the overall analytical approach, and the rationale for each adjustment are discussed in part II.

Summary of Results

The original and adjusted first-year costs are presented in table 14–1 for the high-sulfur-coal scenario and in table 14–2 for the low-sulfur-coal scenario. For both scenarios, coal power is cheaper than nuclear power before and after adjustments. Before adjustments, high-sulfur-coal generation with FGD is 6 percent cheaper than nuclear generation; low-sulfur-coal generation without FGD is 25 percent cheaper than nuclear generation. After adjustments, coal is cheaper than nuclear by 50 percent in the high-sulfur scenario, and by 58 percent in the low-sulfur scenario.

First-year capital costs, before and after adjustments, are cheaper for high- and low-sulfur coal than for nuclear. Before adjustments, high-sulfur coal has a 34-percent cost advantage over nuclear, and low-sulfur coal has a 54-percent cost advantage. After adjustments, the cost advantage increases to 70 percent for high-sulfur coal and 78 percent for low sulfur coal.

First-year fuel costs, before and after adjustments, are higher for both coal scenarios than for nuclear. Before adjustments, high-sulfur coal is 91 percent more expensive than nuclear, and low-sulfur coal is 84 percent more expensive. After adjustments, high-sulfur coal's cost disadvantage decreases slightly to 87 percent, and low-sulfur coal remains unchanged.

Table 14–1
Exxon Case Study, High-Sulfur Coal with FGD: Summary of Original and Adjusted First-Year Busbar Costs for Future Nuclear and Coal Power Units
(mills/kwh)

	Original			Adjusted		
	Nuclear	*Coal*	*Coal as Percentage of Nuclear*	*Nuclear*	*Coal*	*Coal as Percentage of Nuclear*
Capital	40.30	26.50	65.8	75.52	22.80	30.2
Fuel	10.10	19.30	191.1	10.10	18.85	186.6
O&M	2.30	4.00	173.9	5.42	4.00	73.8
Total	52.70	49.80	94.5	91.04	45.65	50.1
CF	60%	66%		55%	70%	

Table 14–2
Exxon Case Study, Low-Sulfur Coal without FGD: Summary of Original and Adjusted First-Year Busbar Costs for Future Nuclear and Coal Power Units
(mills/kwh)

	Original			Adjusted		
	Nuclear	*Coal*	*Coal as Percentage of Nuclear*	*Nuclear*	*Coal*	*Coal as Percentage of Nuclear*
Capital	40.30	18.40	45.7	75.52	16.79	22.2
Fuel	10.10	18.60	184.2	10.10	18.60	184.2
O&M	2.30	2.60	113.0	5.42	2.60	48.0
Total	52.70	39.60	75.1	91.04	37.99	41.7
CF	60%	70%		55%	70%	

First-year O&M costs for high- and low-sulfur coal exceed nuclear's O&M before adjustments; after adjustments, both coal scenarios are cheaper than nuclear. Before adjustments, high-sulfur coal is 74 percent more expensive than nuclear, and low-sulfur coal is 13 percent more expensive. After adjustments, high-sulfur coal is 26 percent cheaper than nuclear and low-sulfur coal is 52 percent cheaper.

The lifetime costs were developed using Exxon's own assumptions. Capital costs are constant over the life of a unit whereas fuel and O&M costs vary. Fuel and O&M costs were escalated, using Exxon's annual

escalation rates, through 12.5 years, which, being the midpoint of a nuclear
unit's lifetime, approximates average lifetime costs. While a coal unit's midpoint is 15 years, all lifetime costs are for 12.5 years of commerciality to
preserve comparability. All O&M costs and coal-fuel costs were escalated at
5 percent per year; nuclear-fuel costs were escalated at 7 percent per year.

The original and adjusted lifetime costs are presented in table 14-3 for
high-sulfur coal and table 14-4 for low-sulfur coal. Lifetime costs follow
patterns similar to first-year costs although coal's cost advantage decreases
with the longer time horizon. Before adjustments, total average costs are
virtually identical for high-sulfur coal and nuclear whereas low-sulfur coal
retains a substantial cost advantage (16 percent) over nuclear. After adjustments, high-sulfur coal power is 41 percent cheaper than nuclear power,
and low-sulfur coal is half as expensive as nuclear. The capital and O&M
lifetime cost ratios are equal or nearly equal to the first-year ratios. The fuel
ratios do vary for lifetime versus first year. High-sulfur coal is 54 percent
more expensive than nuclear fuel before adjustments and 50 percent more
expensive after adjustments. Low-sulfur coal is 48 percent higher than
nuclear both before and after adjustments.

The sensitivity analysis for total average lifetime costs is summarized in
table 14-5. Changing nuclear's capacity factor produces the most significant change in the coal/nuclear total average cost ratio. In the base case of a
55-percent capacity factor for nuclear and 70 percent for coal, high-sulfur
coal is 41 percent cheaper than nuclear, and low-sulfur coal is half as expensive as nuclear. A 50-percent nuclear capacity factor improves coal's cost
advantage by 7 percent over the base case; high-sulfur coal is 46 percent
cheaper than nuclear, and low-sulfur coal is 53 percent cheaper. A
60-percent nuclear capacity factor decreases coal's cost advantage by 7 per-

Table 14-3
**Exxon Case Study, High-Sulfur Coal with FGD: Summary of Original and
Adjusted Lifetime Busbar Costs for Future Nuclear and Coal Power Units**
(mills/kwh)

	Original			Adjusted		
	Nuclear	*Coal*	*Coal as Percentage of Nuclear*	*Nuclear*	*Coal*	*Coal as Percentage of Nuclear*
Capital	40.30	26.50	65.8	75.52	22.80	30.2
Fuel	22.00	33.80	153.6	22.00	33.01	150.0
O&M	4.00	7.00	175.0	9.50	7.00	73.7
Total	66.30	67.30	101.5	107.02	62.81	58.7
CF	60%	66%		55%	70%	

Table 14-4

Exxon Case Study: Low-Sulfur Coal without FGD: Summary of Original and Adjusted Lifetime Busbar Costs for Future Nuclear and Coal Power Units *(mills/kwh)*

	Original			Adjusted		
	Nuclear	*Coal*	*Coal as Percentage of Nuclear*	*Nuclear*	*Coal*	*Coal as Percentage of Nuclear*
Capital	40.30	18.40	45.7	75.52	16.79	22.2
Fuel	22.00	32.60	148.2	22.00	32.60	148.2
O&M	4.00	4.60	115.0	9.50	4.60	48.4
Total	66.30	55.60	83.9	107.02	53.99	50.4
CF	60%	70%		55%	70%	

Table 14-5

Exxon Case Study: Summary of Sensitivity Analysis

Capacity Factor (%)		Total Mills/kwh: Coal as Percentage of Nuclear	
Nuclear	*Coal*	*Hi-S Coal with FGD*	*Low-S Coal without FGD*
55	70	58.7	50.4
60	70	62.8	54.0
50	70	54.4	46.7
55	75	57.3	49.4
55	65	60.3	51.7

cent over the base case; high-sulfur coal is 37 percent cheaper than nuclear, and low-sulfur coal is 46 percent cheaper.

The coal/nuclear total lifetime cost ratios are affected little by changing coal's capacity factor. Raising coal's capacity factor to 75 percent increases coal's cost advantage by 2 percent over the base case; high-sulfur coal is 43 percent cheaper than nuclear, and low-sulfur coal is 51 percent cheaper. Lowering coal's capacity factor to 65 percent decreases coal's cost advantage by 3 percent; high-sulfur coal is 40 percent cheaper than nuclear, and low-sulfur coal is 48 percent cheaper.

The calculations for the lifetime and first-year costs just summarized are detailed in the following sections. Lifetime costs, the focus of the analysis, are addressed first, followed by the sensitivity analysis for lifetime costs. The adjustments to first-year costs are then presented.

Adjustments to Lifetime Costs

(See worksheet in table 14-6.)

1. Technological Sufflation

    a. *Nuclear*
        The 30-percent increase in capital costs is calculated as follows:

$$\text{(Original capital)}/(0.30) = \text{capital increase}$$

$$(40.30)(0.30) = 12.09 \text{ mills/kwh}$$

$$40.30 + 12.09 = 52.39 \text{ mills/kwh} = \text{total adjusted capital}$$

    b. *Coal, High Sulfur*
        No adjustment.

    c. *Coal, Low Sulfur.*
        No adjustment.

2. Construction Time

The Exxon case study provided not the dollar amounts for interest during construction (IDC) and price escalation during construction (EDC) but the percentage of annual capital costs that IDC and EDC constituted. The dollar amounts can be derived using a computation known as the sum of a geometric series given below:

$$(1)\ \ K_1 = \frac{K_T(r-1)}{r(n+1) - r}$$

$$(2)\ \ IDC + EDC = K_T - nK_1$$

where:

k_T = total capital investment, including IDC and EDC, in \$/kw

k_1 = first-year capital cost, excluding IDC and EDC, in \$/kw

r = 1 + (annual percent for IDC and EDC)

n = number of years construction time

Table 14–6
Exxon Case Study: Worksheet for Adjusted Lifetime Busbar Costs of
Future Nuclear and Coal Power Units
(mills/kwh)

		Nuclear	Coal High-Sulfur with FGD	Coal Low-Sulfur without FGD
A.	Original Costs Adjusted for Lifetime			
	Capital	40.30	26.50	18.40
	Fuel	22.00	33.80	32.60
	O&M	4.00	7.00	4.60
	Total	66.30	67.30	55.60
	Capacity factor	60%	66%	70%
B.	Capital Adjustments	55% CF	70% CF	70% CF
	1. Technological sufflation	+ 12.09		
	Adjusted total, capital	52.39		
	2. Construction time	+ 5.30	− 2.32	− 1.61
	Adjusted total, capital	57.69	24.18	16.79
	3. Unit life	+ 11.54		
	Adjusted total, capital	69.23		
	4. Capacity factor	+ 6.29	− 1.38	
	Adjusted total, capital	75.52	22.80	16.79
C.	Fuel Adjustments			
	5. Btu/lb of coal			
	6. Heat rate		− 0.79	
	Adjusted total, fuel	22.00	33.01	32.60
D.	O&M Adjustments			
	7. O&M	+ 5.50		
	Adjusted total, O&M	9.50	7.00	4.60
E.	Other Adjustments			
	8. Yoyo Factor			
	9. Replacement power			
	10. Major repairs			
	11. Senescence			
	12. Fuel prices			
	13. Use of low-sulfur coal			
	14. Waste disposal			
	15. Decommissioning			
F.	Sum of Adjustment[a]	+ 40.72	− 4.49	− 1.61
G.	Adjusted Total Cost[a]	107.02	62.81	53.99
H.	Coal as Percent of Nuclear		58.69	50.45

[a]Adjustments are not included for items 8 through 12, 14, and 15.

$k_1 = k_2 = \ldots k_n$: a linear rate is assumed in the absence of a more specific progression of expenditures

Exxon assumed 7.5 percent for IDC and 5 percent for EDC which gives an $r = 1.125$ for both nuclear and coal.

a. *Nuclear*
    Exxon assumed \$1,210/kw for total capital cost (that is, k_T) and a ten-year construction time (that is, n). Using the preceding formulas, IDC and EDC are calculated as follows:

$$(1)\ K_1 = \frac{1,210(0.125)}{1.125^{11} - 1.125} = \$59.824/kw$$

$$(2)\ IDC + EDC = 1,210 - (10)(59.824) = \$612/kw$$

Having derived Exxon's IDC and EDC, the construction-time adjustment can now be made. Raising nuclear's construction time from ten to twelve years increases IDC and EDC by approximately 20 percent (that is, 2/10). The adjustment is first calculated in dollars per kw and then applied to the mills per kwh number as follows:

$(IDC + EDC$ in \$/kw$)(0.20) = $ \$/kw capital increase

$(\$612)(0.20) = \$122.40/kw$

$(\$/kw$ capital increase$)/(\$/kw$ total capital$) = \%$ increase in total capital

$\$122.4/\$1.210 = 0.1012$

$(\%$ increase in total capital$)($mills/kwh total capital$) = $ mills/kwh capital increase

$(0.1012)(52.39) = 5.30$ mills/kwh

$52.39 + 5.30 = 57.69$ mills/kwh $=$ total adjusted capital

b. *Coal, High Sulfur*
    Exxon assumed \$875/kw for total capital cost (that is, K_T) and a six-year construction time (that is, n). Using the formula for the sum of a geometric series, IDC and EDC are calculated as follows:

$$(1)\ K_1 = \frac{875(0.125)}{1.125^7 - 1.125} = \$94.64/kw$$

(2) IDC + EDC = 875 − (6)(94.64) = $307/kw

Having derived Exxon's IDC and EDC, the construction-time adjustment can now be made. Lowering high-sulfur coal's construction time from 6 to 4.5 years decreases IDC and EDC by approximately 25 percent (that is, 1.5/6.0). The adjustment is first calculated in dollars per kw and then applied to the mills per kwh number as follows:

(IDC + EDC in $/kw)(0.25) = $/kw capital decrease

($307)(0.25) = $76.75/kw

($/kw capital decrease)/($/kw total capital) = % decrease in total capital

$76.75/$875 = 0.0877

(% decrease in total capital)(mills/kwh total capital) = mills/kwh capital decrease

(0.0877)(26.50) = 2.32 mills/kwh

26.50 − 2.32 = 24.18 mills/kwh = total adjusted capital

c. *Coal, Low Sulfur*
Exxon assumed $645/kw for total capital cost (that is, K_T) and a six-year construction time (that is, n). Using the formula for the sum of a geometric series, IDC and EDC are calculated as follows:

(1) $K_1 = \dfrac{645(0.125)}{1.125^7 - 1.125} = \$69.76/kw$

(2) IDC + EDC = 645 − (6)(69.76) = $226/kw

Having derived Exxon's IDC and EDC, the construction-time adjustment can now be made. Reducing low-sulfur coal's construction time from 6 to 4.5 years decreases IDC and EDC by approximately 25 percent (that is, 1.5/6.0). The adjustment is first calculated in dollars per kw and then applied to the mills per kwh number as follows:

(IDC + EDC in $/kw)(0.25) = $/kw capital decrease

($226)(0.25) = $56.50/kw

($/kw capital decrease)/($/kw total capital) = % decrease in total capital

$56.50/645 = 0.0876

(% decrease in total capital)(mills/kwh total capital) = mills/kwh capital decrease

(0.0876)(18.40) = 1.61 mills/kwh

18.40 − 1.61 = 16.79 mills/kwh = total adjusted capital

3. *Unit Life*

   a. *Nuclear*
      Reducing nuclear's unit life from thirty to twenty-five years increases the capital costs by 20 percent (that is, 5/25) as follows:

      (Total capital)(0.20) = capital increase

      (57.69)(0.20) = 11.54 mills/kwh

      57.69 + 11.54 = 69.23 mills/kwh = total adjusted capital

   b. *Coal, High Sulfur*
      No adjustment.

   c. *Coal, Low Sulfur*
      No adjustment.

4. *Capacity Factor*

   a. *Nuclear*
      Lowering nuclear's capacity factor from 60 to 55 percent increases capital costs by 9.09 percent (that is, 5/55) as follows:

      (Total capital)(0.0909) = capital increase

      (69.23)(0.0909) = 6.29 mills/kwh

      69.23 + 6.29 = 75.52 mills/kwh = total adjusted capital

   b. *Coal, High Sulfur*
      Increasing high-sulfur coal's capacity factor from 66 to 70 percent decreases capital costs by 5.71 percent (that is, 4/70) as follows:

(Total capital)(0.0571) = capital increase

(24.18)(0.0571) = 1.38 mills/kwh

24.18 − 1.38 = 22.80 mills/kwh = total adjusted capital

c. *Coal, Low Sulfur*
No adjustment was necessary as Exxon assumed a 70-percent capacity factor for low-sulfur coal.

5. *Btu per Pound of Coal*

a. *Nuclear*
No adjustment.

b. *Coal, High Sulfur*
No adjustment was possible as Exxon expressed fuel costs in ¢/million Btu.

c. *Coal, Low Sulfur*
No adjustment was possible as Exxon expressed fuel costs in ¢/million Btu.

6. *Heat Rate*

a. *Nuclear*
No adjustment.

b. *Coal, High Sulfur*
Reducing Exxon's assumed heat rate for high-sulfur coal from 9,420 to 9,200 Btu/kwh decreases coal's fuel costs by 2.34 percent (that is, 220/9,420) as follows:

(Original fuel)(0.0234) = fuel decrease

(33.80)(0.0234) = 0.79 mills/kwh

33.80 − 0.79 = 33.01 mills/kwh = total adjusted fuel

c. *Coal, Low Sulfur*
No adjustment was necessary as Exxon assumed 9,050 Btu/kwh for low-sulfur coal.

7. *O&M*

   a. *Nuclear*
      The 1978 average O&M for large nuclear units (3.85 mills/kwh with an average capacity factor of 55 percent) was escalated at Exxon's annual rate of 5 percent through 12.5 years of commerciality. The result is a lifetime O&M of 9.50 mills/kwh, or a 5.50-mill increase over the base cost of 4.00 mills/kwh.

   b. *Coal, High Sulfur*
      No adjustment.

   c. *Coal, Low Sulfur*
      No adjustment.

Sensitivity Analysis for Lifetime Costs

Nuclear Capacity Factor

Changing nuclear's capacity factor affects capital and O&M costs by increasing or decreasing the number of kwh over which the costs are averaged. The sensitivity results for the 60-percent and 50-percent nuclear capacity factors are presented in table 14–7 for the high-sulfur-coal scenario and table 14–8 for the low-sulfur-coal scenario. While nuclear's costs are the same across the two scenarios, coal as a percent of nuclear, of course, differs. The calculations are outlined as follows.

Table 14–7
Exxon Case Study, High-Sulfur Coal with FGD: Sensitivity
Analysis on Nuclear Capacity Factor

	Nuclear Costs (mills/kwh)			Coal as Percentage of Nuclear[a]		
	N = 55%	N = 60%	N = 50%	N = 55%	N = 60%	N = 50%
Capital	75.52	69.23	83.08	30.2	32.9	27.4
Fuel	22.00	22.00	22.00	150.0	150.0	150.0
O&M	9.50	8.71	10.45	73.7	80.4	67.0
Total	107.02	99.94	115.53	58.7	62.8	54.4

[a]Assumes a 70-percent capacity factor for coal.

Table 14–8
**Exxon Case Study, Low-Sulfur Coal without FGD: Sensitivity
Analysis on Nuclear Capacity Factor**

	Nuclear Costs (mills/kwh)			Coal as Percentage of Nuclear[a]		
	N = 55%	N = 60%	N = 50%	N = 55%	N = 60%	N = 50%
Capital	75.52	69.23	83.08	22.2	24.3	20.2
Fuel	22.00	22.00	22.00	148.2	148.2	148.2
O&M	9.50	8.71	10.45	48.4	52.8	44.0
Total	107.02	99.94	115.53	50.4	54.0	46.7

[a]Assumes a 70-percent capacity factor for coal.

Capital Costs. Adjustment 4 on the worksheet in table 14–6 captures the capacity-factor change for capital costs. A 60-percent capacity factor was originally assumed by Exxon so no adjustment is necessary for the $N = 60$ percent case. A 50-percent capacity factor increases capital costs by 20 percent over the original Exxon number (that is, 10/50) as follows:

$$(\text{Total capital})(0.20) = \text{capital increase}$$

$$(69.23)(0.20) = 13.85 \text{ mills/kwh}$$

$$69.23 + 13.85 = 83.08 \text{ mills/kwh} = \text{total adjusted capital}$$

It should be noted that, relative to the 55-percent base case, the $N = 60$ percent case decreased capital costs and the $N = 50$ percent case increased capital costs.

O&M Costs. The capacity-factor adjustment for O&M costs is made at item 7 on the table 14–6 worksheet. The 60-percent capacity factor decreases the base-case O&M but increases Exxon's original O&M as follows:

$$(55\% \text{ CF O\&M})(55/60) = \text{total adjusted O\&M}$$

$$(9.50)(0.9167) = 8.71 \text{ mills/kwh}$$

$$(\text{Total adjusted O\&M}) - (\text{original O\&M}) = \text{O\&M increase}$$

$$8.71 - 4.00 = 4.71 \text{ mills/kwh}$$

The 50-percent capacity factor increases both the base case and Exxon's original O&M as follows:

$(55\%$ CF O&M)$(55/50)$ = total adjusted O&M

$(9.50)(1.10)$ = 10.45 mills/kwh

(Total adjusted O&M) $-$ (original O&M) = O&M increase

$10.45 - 4.00$ = 6.45 mills/kwh

Coal Capacity Factor

Changing coal's capacity factor affects capital costs only by reducing the number of kwh over which costs are averaged (item 4 on the table 14–6 worksheet). The results are presented in table 14–9 for the high-sulfur-coal scenario and table 14–10 for the low-sulfur-coal scenario. The calculations are explained as follows.

Capital Costs. Raising high-sulfur coal's capacity factor to 75 percent reduces capital costs by 12.0 percent over Exxon's original assumption of 66 percent (that is, 9/75) as follows:

(Total capital)(0.1200) = capital decrease

$(24.18)(0.1200)$ = 2.90 mills/kwh

$24.18 - 2.90$ = 21.28 mills/kwh = total adjusted capital

Raising low-sulfur coal's capacity factor to 75 percent reduces capital costs by 6.67 percent over Exxon's original case (that is, 5/75) as follows:

(Total capital)(0.0667) = capital decrease

$(16.79)(0.0667)$ = 1.12 mills/kwh

Table 14–9
Exxon Case Study: High-Sulfur Coal with FGD: Sensitivity
Analysis on Coal Capacity Factor

| | Coal Costs (mills/kwh) | | | Coal as Percentage of Nuclear[a] | | |
	$C = 70\%$	$C = 75\%$	$C = 65\%$	$C = 70\%$	$C = 75\%$	$C = 65\%$
Capital	22.80	21.28	24.55	30.2	28.2	32.5
Fuel	33.01	33.01	33.01	150.0	150.0	150.0
O&M	7.00	7.00	7.00	73.7	73.7	73.7
Total	62.81	61.29	64.56	58.7	57.3	60.3

[a]Assumes a 55-percent capacity factor for nuclear.

Table 14–10
**Exxon Case Study: Low-Sulfur Coal without FGD: Sensitivity
Analysis on Coal Capacity Factor**

	Coal Costs (mills/kwh)			Coal as Percentage of Nuclear[a]		
	C = 70%	C = 75%	C = 65%	C = 70%	C = 75%	C = 65%
Capital	16.79	15.67	18.08	22.2	20.7	23.9
Fuel	32.60	32.60	32.60	148.2	148.2	148.2
O&M	4.60	4.60	4.60	48.4	48.4	48.4
Total	53.99	52.87	55.28	50.4	49.4	51.7

[a]Assumes a 55-percent capacity factor for nuclear.

$$16.79 - 1.12 = 15.67 \text{ mills/kwh} = \text{total adjusted capital}$$

A 65-percent capacity factor for high-sulfur coal increases capital costs by 1.54 percent over Exxon's original case (that is, 1/65) as follows:

$$\text{(Total capital)}(0.0154) = \text{capital increase}$$

$$(24.18)(0.0154) = 0.37 \text{ mills/kwh}$$

$$24.18 + 0.37 = 24.55 \text{ mills/kwh} = \text{total adjusted capital}$$

Reducing low-sulfur coal's capacity factor to 65 percent increases capital costs by 7.69 percent over Exxon's original case (that is, 5/65) as follows:

$$\text{(Total capital)}(0.0769) = \text{capital increase}$$

$$(16.79)(0.0769) = 1.29 \text{ mills/kwh}$$

$$16.79 + 1.29 = 18.08 \text{ mills/kwh} = \text{total adjusted capital}$$

Adjustments to First-Year Costs

The adjustments outlined in part II were applied to Exxon's first-year costs as well as to lifetime costs. The capital-cost adjustments are identical for the first year and lifetime, and they are not therefore outlined again here. The fuel and O&M adjustments are conceptually the same for lifetime and first year of commerciality but are applied on a lower base for the latter. The worksheet for the adjusted first-year costs is contained in table 14–11 and the fuel and O&M adjustments are explained as follows.

Table 14–11
Exxon Case Study: Worksheet for Adjusted First-Year Busbar Costs of Future Nuclear and Coal Power Units
(mills/kwh)

		Nuclear	Coal High-Sulfur with FGD	Coal Low-Sulfur without FGD
A.	Original Costs			
	Capital	40.30	26.50	18.40
	Fuel	10.10	19.30	18.60
	O&M	2.30	4.00	2.60
	Total	52.70	49.80	39.60
	Capacity factor	60%	66%	70%
B.	Capital Adjustments	55% CF	70% CF	70% CF
	1. Technological sufflation	+ 12.09		
	Adjusted total, capital	52.39		
	2. Construction time	+ 5.30	− 2.32	− 1.61
	Adjusted total, capital	57.69	24.18	16.79
	3. Unit life	+ 11.54		
	Adjusted total, capital	69.23		
	4. Capacity factor	+ 6.29	− 1.38	
	Adjusted total, capital	75.52	22.80	16.79
C.	Fuel Adjustments			
	5. Btu/lb of coal			
	6. Heat rate		− 0.45	
	Adjusted total, fuel	10.10	18.85	18.60
D.	O&M Adjustments			
	7. O&M	+ 3.12		
	Adjusted total, O&M	5.42	4.00	2.60
E.	Other Adjustments			
	8. Yoyo Factor			
	9. Replacement power			
	10. Major repairs			
	11. Senescence			
	12. Fuel prices			
	13. Use of low-sulfur coal			
	14. Waste disposal			
	15. Decommissioning			
F.	Sum of Adjustment[a]	+ 38.34	− 4.15	− 1.61
G.	Adjusted Total Cost[a]	91.04	45.65	37.99
H.	Coal as Percent of Nuclear		50.14	41.73

[a]Adjustments are not included for items 8 through 12, 14, and 15.

1. through 4.   Capital Adjustments

Same as for lifetime costs.

5.   Btu per Pound of Coal

    a.   *Nuclear*
        No adjustment.

    b.   *Coal, High Sulfur*
        No adjustment was possible as Exxon expressed fuel costs in ¢/million Btu.

    c.   *Coal, Low Sulfur*
        No adjustment was possible as Exxon expressed fuel costs in ¢/million Btu.

6.   Heat Rate

    a.   *Nuclear*
        No adjustment.

    b.   *Coal, High Sulfur*
        Reducing Exxon's assumed heat rate for high-sulfur coal from 9,420 to 9,200 Btu/kwh decreases coal's fuel costs by 2.34 percent (that is, 220/9,420) as follows:

$$(\text{Original fuel})(0.0234) = \text{fuel decrease}$$

$$(19.30)(0.0234) = 0.45 \text{ mills/kwh}$$

$$19.30 - 0.45 = 18.85 \text{ mills/kwh} = \text{total adjusted fuel}$$

    c.   *Coal, Low Sulfur*
        No adjustment was necessary as Exxon assumed 9,050 Btu/kwh for low-sulfur coal.

7.   O&M

    a.   *Nuclear*
        The 1978 average O&M for large nuclear units (3.85 mills/kwh with an average capacity factor of 55 percent) was escalated at Exxon's

annual rate of 5 percent through the first year of commerciality. The result is a first year O&M of 5.42 mills/kwh, or a 3.12-mill increase over the original cost of 2.30 mills/kwh.

b. *Coal, High Sulfur*
    No adjustment.

c. *Coal, Low Sulfur*
    No adjustment.

15 Summary and Conclusions

The four case studies, adjusted for realistic assumptions, indicate that the total average lifetime busbar costs (mills/kwh) for coal are 22 to 50 percent cheaper than for nuclear. In the two most important case studies of high-sulfur coal with FGD—NRC and Exxon—coal is 41 and 44 percent cheaper than nuclear, respectively. In the low-sulfur coal without FGD scenario, coal is half as expensive as nuclear.

The case-study results are conservative in favor of nuclear power. The 30-percent technological-sufflation adjustment for nuclear probably underestimates nuclear's capital costs. Further, the results include zero or insignificant amounts for the Yoyo Factor, major repairs, replacement power, senescence, waste disposal, and decommissioning.

The case studies do not include another component of nuclear power's costs, namely, federal subsidies. A draft report by the U.S. Department of Energy (December 1980) estimated that federal subsidies to the nuclear power industry over the past thirty years totalled $37 billion. The report claimed that, without these subsidies, nuclear-power generating costs would be 150 to 200 percent higher. The final revised version reduced these subsidies to $12.8 billion by excluding certain research on military reactors, half of enrichment subsidies, and $6.5 billion of research on nuclear-waste disposal that has not yet been translated into industrial use (U.S. Department of Energy, *Federal Support for Nuclear Power: Reactor Design and Fuel Cycle,* DOE/EIA–0201/13, Final Report, 1982).

Nuclear power's cost disadvantage relative to coal is attributable to nuclear's design/performance gap. Total costs are higher than expected (for example, initial capital costs, major postconstruction repairs, and O&M costs), while the kwh over which costs are averaged are fewer than expected. The design/performance gap results from design-technological risks (for example, scaling) and human factors. An analysis of learning in the nuclear power industry indicates that the design/performance gap has not been narrowing and may, in fact, be increasing for a portion of the industry.

Several aspects of the competitive economics of nuclear and coal power deserve further treatment and refinement. Data on the nuclear power industry need "sanitizing": definitive commerciality dates, capacity ratings, and capacity factors, for example, should be ascertained and published. Similarly, accurate unit (as opposed to plant) data for the coal power industry should be publicly available. Credible quantification of nuclear capital costs, the Yoyo Factor, major repairs, replacement power, waste disposal, and decommissioning is essential. The fact that such crucial cost components have not been satisfactorily quantified reflects the primary difficulties in the nuclear power industry—inadequate available data and the technological and economic gap between design and actual performance.

Appendix:
Safety, Financial, and Data Aspects of Commerciality Dates for U.S. Nuclear Power Units

The purpose of this appendix is threefold:

To identify two federal-income-tax policies that may promote premature commerciality of nuclear generating units, and to discuss the safety and financial implications of premature commerciality for the nuclear power industry

To identify problems with commerciality data in the public record that may or may not stem from the federal-income-tax policies

To examine whether premature commerciality may, in fact, have occurred in the U.S. nuclear power industry

Each objective is addressed in separate sections that follow.

Premature Commerciality: Incentives and Implications

The commercial date of a generating unit represents a significant financial milestone for a utility. With few exceptions, commerciality determines when and how a unit's costs are recognized for federal income taxes and rate making. Until 1979 the only precommerciality expenses deductible for taxes were interest on borrowed funds; a percentage of allowed investment-tax credits for construction payments (phased-in by 20-percent increments from 1975 to 1979); and property taxes. Since 1979, full investment tax credits (ITCs) on construction payments may also be taken before commerciality as progress payments. Except for the possible allowance of construction work in progress (CWIP) in the rate base, no costs can be recouped in rates until commerciality, at which time all precommerciality costs are capitalized.

The specific tax and rate-making benefits that begin at commerciality are as follows:

159

Federal-income-tax benefits

   Deducting depreciation

   Deducting operating and maintenance (O&M) expenses

   Receiving full ITCs (until 1979, available only in part before commerciality)

Federal and state rate-making benefits

   Including precommerciality costs in the rate base

   Receiving O&M costs

   Expensing depreciation

It is important to note that commerciality determines not only when costs are recognized for taxes and rate-making but how they are recognized. Costs incurred before commerciality are generally capitalized and recouped over time through depreciation. Many costs incurred after commerciality are treated as operating-and-maintenance expenses, which in their entirety are passed on in rates and deducted from taxable income. Commerciality dates are used to determine when and how a unit's costs are recognized for state income taxes as well.

The focus of this appendix is federal-income-tax treatment of depreciation and ITCs. Until 1979, two retroactive policies created unusual pressures for premature commerciality. The first, known as the "six-month rule," allowed depreciation for federal income taxes back to January 1 and July 1 if a unit was declared commercial by June 30 and December 31, respectively. The second, to be called here the "twelve-month rule," allowed full investment tax credits back to January 1 regardless of when in the calendar year a unit went commercial. The six- and twelve-month rules made December 31 and January 1 very different commerciality dates from a financial point of view. The six-month rule also made June 30 and July 1 significantly different commerciality dates.

Recent changes in the tax laws have altered but not eliminated incentives for premature commerciality. Since 1981, the six-month rule has been modified—six months of depreciation are allowed for the first year of commerciality regardless of when in the calendar year a unit is declared commercial. Thus, June 30 has lost its financial significance; December 31 versus January 1 commerciality still retains its financial advantage (that is, six extra months of depreciation on a very large investment).

The 1979 change in the ITC laws (allowing full progress payments during construction) lessens but does not obviate the incentive for premature commerciality. A December 31 commerciality date allows the ITC to be taken in full for that tax year. A January 1 (or later) commerciality date spreads the ITC benefit over the four quarterly estimated tax payments. A further incentive exists to declare a unit commercial on December 31 versus January 1. Estimated taxes are based on the previous year's actual income or the current year's estimated income, whichever is lower. ITCs and depreciation taken as of December 31 lower the estimated income and, hence, the required estimated taxes for the following year.

Two potential incentives exist for postponing commerciality. First, property taxes in some locales may increase substantially at commerciality. In such cases, a January 1 commerciality date would be preferable to a December 31 date. Second, ITC carry-over rules may encourage a utility to declare commerciality as late as possible. The utility may have accumulated a large backlog of ITCs to carry over to future tax years. January 1 versus December 31 commerciality allows the utility one more year in the carry-over period (seven years until 1981 and 15 years since 1981) for that ITC.

Incentives for declaring commerciality are not problematic in and of themselves. If the relevant tax and rate-making policies clearly state and strictly enforce definitions of commerciality, it would be difficult for a unit to be declared commercial before meeting the specified criteria. If, however, commerciality is not clearly defined in the relevant policies, two types of premature commerciality may result. The first type—premature accounting commerciality—involves a declaration of commerciality (and hence the receipt of financial benefits) while a unit is not operating at truly commercial levels. The second type—premature operating commerciality—involves a declaration of commerciality (and hence the receipt of financial benefits) and operations at commercial levels without the demonstrated capability to operate at sustained normal levels. Operating commerciality could have several outcomes. At best, the unit is not significantly affected by the prematurely high levels of operation. However, if the premature unit develops problems, either the worst scenario ensues, that is, a nuclear accident, or the unit falls to operating levels that place it in the accounting commerciality category. In short, accounting and in some cases operating commerciality result in premature financial benefits; operating commerciality may also involve much more serious safety hazards.

The financial implications of premature commerciality, summarized in table A-1, are numerous. Incorrect depreciation and O&M expenses make taxes and rates incorrect. Incorrect ITCs affect taxes; an incorrect rate base as well as incorrect taxes, O&M, and depreciation affect rates. In addition, accounting commerciality would affect economic studies of electricity generation by creating incorrect total and average lifetime costs. Total lifetime O&M and capital costs would be incorrect because some of the O&M costs should have been capitalized as precommercialization costs. Average O&M and capital costs would be incorrect both because the total costs are incorrect (that is, the numerator) and because the age of plant and hence the capacity factor is incorrect (that is, the denominator).

Table A-1
Financial and Economic Results of Accounting Commerciality

	Affected Activities		
Result of Accounting Commerciality	Income Taxes	Rate Making	Economic Analyses
1. Incorrect depreciation	X	X	
2. Incorrect investment-tax credits	X		
3. Incorrect operating-and-maintenance costs	X	X	
4. Incorrect rate base		X	
5. Incorrect average lifetime operating-and-maintenance and capital costs			X

To summarize thus far, financial benefits provide the incentive for premature commerciality, while ambiguous tax and rate-making policies provide the opportunity. Premature commerciality in turn creates financial and potential safety problems for nuclear units. Premature commerciality is important for fossil-fuel units as well. Unfortunately, data are not published for individual fossil-fuel units (only for plants) so that analysis is not feasible. The accounting problems of premature commerciality do not differ for fossil versus nuclear units; the potential safety problems, however, generally apply to nuclear units only.

Before examining whether the six- and twelve-month rules may have, in fact, promoted premature commerciality, problems with public data on commerciality need to be addressed.

Data Problems

Commerciality dates in the public record are in a confused state. Prairie Island-2 and Palisades, for example, went commercial on the same day that first electricity was achieved, an engineering and definitional impossibility. Different sources, and even the same source, often contain conflicting commerciality dates for the same unit. The confusion could be due to several factors. Some of the published dates may simply be incorrect. Alternatively, particular units may in fact have been given multiple commerciality dates by the owner. It may well be financially advantageous for a utility to use different commerciality dates for different purposes, for example, for taxes, rate making, real-estate property taxes, and acceptance into a power pool.

If multiple commerciality dates are legitimate, the public record should provide a clear explanation of the various dates. However, tax and rate-making policies that do not clearly define commerciality create the opportunity for multiple commerciality dates. If at least one of the dates constitutes premature commerciality, the safety and financial problems discussed previously may result.

Commerciality dates for seventy U.S. nuclear units were compiled for this study from some or all of the following sources:

NRC's monthly *Status Summary Report* and *Annual Report*

FERC/FPC, *Statistics of Steam Electric Plants*

FERC/FPC staff

International Atomic Energy Agency's (IAEA) annual *Operating Experience with Nuclear Power Stations*

Utility annual financial reports

State public-utility commissions

For most units, commerciality dates were taken from the first three sources. Further sources were consulted if the first three indicated discrepancies.

Forty-one percent of the seventy units had conflicting commerciality dates in the various sources. Thirty percent of all units had two conflicting dates, 8 percent had three conflicting dates, and 3 percent had more than three conflicting dates. Twenty percent of the seventy units had significant conflicts (that is, greater than one month's difference) across sources. Those units, as well as their conflicting commerciality dates, are listed in table A–2.

Evidence for Premature Commerciality

Three tests for premature commerciality were used. First, were an unusual number of units declared commercial in what might be called the "pressure periods," that is, the last half of June and December? Second, did units that went commercial in a pressure period move more quickly from criticality to commerciality than other units? Third, did units that went commercial in a pressure period have lower capacity factors in the first twelve months of commercial operation than other units? It should be noted that, if a unit had multiple commerciality dates, one of which fell in a pressure period, the latter was used in the analysis. The logic is that the pressure-period date was used for some purpose.

These three premature commerciality tests were applied to sixty-three

Table A-2
U.S. Nuclear Units with Significant Discrepancies in Published
Commerciality Dates

Unit Name	Published Commerciality Dates (and Sources)	
1. Big Rock Point-1	8-31-62	(FPC staff)
	12-62	(FPC *Statistics*)
	3-29-63	(NRC monthly *Status Report*)
2. Dresden-2	7-70	(NRC monthly *Status Report*)
	8-11-70	(FPC staff and *Statistics*)
	1-71	(IAEA *Operating Experience*)
	6-9-72	(NRC *Annual Report*)
3. Duane Arnold	6-1-74	(FPC staff)
	6-22-74	(FPC *Statistics*)
	2-1-75	(NRC monthly *Status Report* and IAEA *Operating Experience*)
4. Ft. Calhoun-1	9-26-73	(FPC staff and *Statistics*)
	6-20-74	(NRC monthly *Status Report*)
5. Ft. St. Vrain	1-1-79	(FPC staff)
	7-1-79	(NRC monthly *Status Report*)
6. Ginna	3-70	(NRC *Annual Report* and IAEA *Operating Experience*)
	7-1-70	(FPC staff, NRC monthly *Status Report,* and N.Y. Public Service Commission)
7. Humbolt Bay	6-63	(FPC *Statistics*)
	8-63	(NRC monthly *Status Report*)
8. Millstone-1	12-28-70	(FPC staff and *Statistics*)
	3-71	(NRC monthly *Status Report*)
9. Palisades	11-71	(FPC *Statistics*)
	12-31-71	(NRC monthly *Status Report*)
	4-15-72	(FPC staff)
10. Quad Cities-1	8-16-72	(FPC staff and *Statistics*)
	2-18-73	(NRC monthly *Status Report*)
11. Quad Cities-2	9-72	(FPC *Statistics*)
	10-24-72	(FPC staff)
	3-10-73	(NRC monthly *Status Report*)
12. San Onofre-1	8-1-67	(FPC staff)
	1-1-68	(FPC *Statistics* and NRC monthly *Status Report*)
13. Trojan	12-24-75	(NRC monthly *Status Report* and IAEA *Operating Experience*)
	5-14-76	(FPC staff)
	5-20-76	(FPC *Statistics* and NRC monthly *Status Report*)
14. Zion-1	10-2-73	(FPC staff and *Statistics*)
	12-31-73	(NRC monthly *Status Report*)

U.S. nuclear units with nameplate ratings of 500 MW or greater. All sixty-three units went commercial before 1981 and were subject, therefore, to the six-month rule for depreciation. Sixty of the units went commercial before 1979 when the twelve-month rule for ITCs was in effect. The remaining three units went commercial in 1979 and 1980. The data (and their sources) used for the premature-commerciality tests are presented at the end of this appendix (tables A-4, A-5, and A-6). The results for each test are discussed separately in the following sections and summarized in table A-3.

Incidence of Commerciality in Pressure Periods

On a purely stochastic basis, approximately five (31/365 days or 8.5 percent) of the sixty-three units would have gone commercial in the last half of June (6-16 through 6-30) and December (12-16 through 12-31). In fact, twenty-two (34.9 percent) of the sixty-three units went commercial in the pressure periods. Moreover, not a single publicly owned unit, to which the six- and twelve-month rules did not apply, went commercial in the pressure periods. Thus the more significant statistic is that, of the fifty-five private

Table A-3
Results of Premature Commerciality Tests

	Pressure Period Units (1)	Other Units (2)	(2) ÷ (1)
Commerciality Dates			
Public and Private Units (63)			
Number of units	22	41	—
Expected percentage	8.5%	91.5%	—
Actual percentage	35.5%	64.5%	—
Private Units (55)			
Number of units	22	33	—
Expected percentage	8.5%	91.5%	—
Actual percentage	40.0%	59.3%	—
Elapsed Days			
(63 public and private units)			
Mean	146	185	+ 26.7%
Median	136	176	+ 29.4%
First 12 Months' Capacity Factor			
(63 public and private units)			
Mean	47.7%	54.5%	+ 14.3%

units subject to the six- and twelve-month rules, 40.0 percent went commercial in the pressure periods.

Looking at the December "double-pressure" period only (significant both for depreciation and ITCs), approximately two (16/365 days or 4.4 percent) of the fifty-five privately owned units would have gone commercial in the last half of December on a stochastic basis. In fact, eighteen (or 32.7 percent) of the units went commercial in this period.

Time from Criticality to Commerciality

Having established that a disproportionately large number of units were declared commercial in the pressure periods, the next step examines whether those units moved on average more quickly from initial criticality to commerciality (called here *elapsed days*) than other units. The number of days from criticality to commerciality averaged 146 for the twenty-two pressure-period units versus 185 for the forty-one other units. Thus the forty-one non-pressure-period units took, on average, 26.7 percent longer to reach commerciality. The median elapsed days for the non-pressure-period units (176 days) was 29.4 percent higher than for the pressure-period units (136 days).

First Twelve Months' Performance

Having established that the pressure-period units declared commerciality more quickly than other units, the final question is whether operating performance (as measured by nameplate capacity factor) in the first twelve months of commercial operation differed for pressure-period versus other units.

The first twelve months' capacity factor averaged 47.7 percent for the twenty-two pressure-period units and 54.5 percent for the forty-one other units. Thus capacity factors for the forty-one units averaged 14.3 percent higher than for the twenty-two pressure-period units.

It is difficult to determine whether a low capacity factor for the first twelve months of commercial operation stems from accounting commerciality, operating commerciality, or neither. However, an examination of a unit's operations during the first few months of commerciality provides some interesting hints. For example, Three Mile Island-2 (TMI-2) went commercial on December 30, 1978, which is within a pressure period. TMI-2 may have been a case of premature operating commerciality—the unit operated at commercial levels for three months until its accident and indefinite shutdown on March 28, 1979.

Capacity-factor data for a number of units suggest premature-accounting commerciality. Zion-1, Trojan, and Palisades were declared commercial during pressure periods (December 31, 1973, December 24, 1975, and December 31, 1971, respectively). Zion-1 did not operate at all, and Trojan and Palisades barely operated during the first three months of commerciality. Peach Bottom-3, declared commercial on December 23, 1974, exhibited a similar pattern for the first two months of commerciality.

Conclusions

Based on the three criteria just examined, premature commerciality may be a frequent occurrence in the nuclear power industry. Federal and state regulations and their enforcement should be reexamined and revised to eliminate, to the extent feasible, incentives for premature commerciality.

Table A–4
Privately Owned Units Declared Commercial in Pressure Periods

Unit Name	Operating Utility	Nameplate (MW)	Date Critical	Date Commercial	Elapsed Days	First 12 Months Capacity Factor (%)
1. Arkansas-1	Arkansas P&L	903	8- 6-74	12-19-74	135	61.6
2. Davis-Besse-1	Toledo Ed. C.	962	8-12-77	12-31-77	141	30.9
3. Duane Arnold	Iowa El. L&P	597	3-23-74	6-22-74	91	45.3
4. Hatch-1	Georgia PC	850	9-12-74	12-31-75	475	55.1
5. Kewaunee	Wisconsin PS	560	3- 7-74	6-22-74	107	60.8
6. Maine Yankee	Maine Yankee Atomic PC	864	10-23-72	12-28-72	66	44.3
7. Millstone-1	Northeast Uts.	690	10-26-70	12-28-70	63	59.2
8. Millstone-2	Northeast Uts.	910	10-17-75	12-26-75	70	56.1
9. Monticello	Northern States PC	569	12-10-70	6-30-71	202	56.1
10. Oconee-3	Duke PC	934	9- 5-74	12-16-74	102	60.7
11. Oyster Creek-1	Jersey Central P&L	670	5- 3-69	12-23-69	234	58.2
12. Palisades	Consumers PC	812	5-24-71	12-31-71	221	25.3

		Nameplate MW	Date critical	Date commercial		Capacity factor
13. Peach Bottom-3	Philadelphia El. C.	1,152	8- 7-74	12-23-74	138	52.7
14. Point Beach-1	Wisconsin El. P.	524	11- 2-70	12-21-70	49	71.0
15. Prairie Island-1	Northern States PC	593	12- 1-73	12-16-73	15	28.0
16. Prairie Island-2	Northern States PC	593	12-17-74	12-21-74	4	59.2
17. Salem-1	Public Service El.&G	1,170	12-11-76	6-30-77	201	36.4
18. St. Lucie-1	Florida P&L	850	4-22-76	12-21-76	243	70.7
19. Surry-1	Virginia El. PC	848	7- 1-72	12-22-72	174	47.1
20. Three Mile Island-2	Metropolitan Ed. C.	961	3-28-78	12-30-78	277	15.7
21. Trojan	Portland G.E.	1,216	12-15-75	12-24-75	9	18.9
22. Zion-1	Commonwealth Ed. C.	1,098	6-19-73	12-31-73	195	36.3

Source: Nameplate MW: Nuclear Regulatory Commission (NRC) staff in correspondence, except for Maine Yankee (provided by New England Electric Company), Millstone-1 (provided by Northeast Utilities), and Oyster Creek-1 (from the Atomic Energy Commission). Date critical: NRC's *Operating Units Status Report*, November 1979. Date commercial: FERC staff (who spoke, in many cases, with the relevant utility); FERC's *Statistics of Steam-Electric Plants*; NRC's *Operating Units Status Report*; NRC's *Nuclear Power Plant Operating Experience*; IAEA's *Operating Experience*; and the 1972 *Annual Report* for Maine Yankee Atomic Power Company (for Maine Yankee only). Capacity factor: Calculated from the sources cited for commerciality dates.

Table A-5
Privately Owned Units Not Declared Commercial in Pressure Periods

Unit Name	Operating Utility	Nameplate (MW)	Date Critical	Date Commercial	Elapsed Days	First 12 Months Capacity Factor (%)
1. Arkansas-2	Arkansas P&L	959	12- 5-78	3-26-80	477	61.4
2. Beaver Valley-1	Duquesne LC	923	5-10-76	9-30-76	143	28.7
3. Brunswick-1	Carolina P&L	867	10- 8-76	3-18-77	161	46.6
4. Brunswick-2	Carolina P&L	867	3-20-75	11- 3-75	228	38.9
5. Calvert Cliffs-1	Baltimore G&E	918	10- 7-74	5- 8-75	213	70.7
6. Calvert Cliffs-2	Baltimore G&E	911	11-30-76	4- 1-77	122	79.2
7. Cook-1	American El. PC	1,152	1-18-75	8-27-75	221	62.7
8. Cook-2	American El. PC	1,133	3-10-78	7- 1-78	113	64.8
9. Crystal River-3	Florida PC	890	1-14-77	3-13-77	58	65.7
10. Dresden-2	Commonwealth Ed. C	828	1- 7-70	8-11-70	216	29.5
11. Dresden-3	Commonwealth Ed. C	828	1-31-71	10-31-71	273	64.8
12. Farley-1	Alabama PC	888	8- 9-77	12- 1-77	114	73.9
13. Ginna	Rochester G&E	517	11- 8-69	7- 1-70	235	54.6
14. Haddam Neck	Conn. Yankee Atomic PC	600	7-24-67	1- 1-68	161	57.1
15. Hatch-1	Georgia PC	809	7- 4-78	9- 5-79	428	60.6
16. Indian Point-2	Consolidated Edison Co.	1,013	5-22-73	8-15-73	85	17.5

17. Nine Mile Point-1	Niagara Mohawk PC	642	9- 5-69	12-14-69	100	32.4
18. North Anna-1	Virginia El. PC	980	4- 5-78	6- 6-78	62	68.4
19. North Anna-2	Virginia El. PC	947	6-12-80	12-14-80	185	66.1
20. Oconee-1	Duke PC	934	4-19-73	7-16-73	88	54.3
21. Oconee-2	Duke PC	934	11-11-73	9- 9-74	302	56.0
22. Peach Bottom-2	Philadelphia El. C	1,152	9-16-73	7- 5-74	292	65.7
23. Pilgrim-1	Boston Ed. C	678	6-16-72	12- 9-72	176	68.6
24. Point Beach-2	Wisconsin El. PC	524	5-30-72	10- 1-72	124	45.9
25. Quad Cities-1	Commonwealth Ed. C	828	10-18-71	8-16-72	303	67.4
26. Quad Cities-2	Commonwealth Ed. C	828	4-26-72	10-24-72	181	69.6
27. Robinson-2	Carolina PL	769	9-20-70	3- 7-71	168	46.8
28. Surry-2	Virginia El. PC	848	3- 7-73	4- 3-73	27	64.5
29. Three Mile Island-1	Metropolitan Ed. C	871	6- 5-74	9- 2-74	89	77.0
30. Turkey Point-3	Florida P&L	760	10-20-72	12- 4-72	45	49.7
31. Turkey Point-4	Florida P&L	760	6-11-73	9- 7-73	88	61.3
32. Vt. Yankee-1	Vermont Yankee Nuclear PC	563	3-24-72	11-30-72	251	37.0
33. Zion-2	Commonwealth Ed. C	1,098	12-24-73	9-19-74	269	42.7

Sources: See table A–4.

Table A-6
Publicly Owned Units

Unit Name	Operating Utility	Nameplate (MW)	Date Critical	Date Commercial	Elapsed Days	First 12 Months Capacity Factor (%)
1. Browns Ferry-1	Tennessee Valley A.	1,152	8-17-73	8- 1-74	349	38.8
2. Browns Ferry-2	Tennessee Valley A.	1,152	7-20-74	3- 1-75	224	16.3
3. Browns Ferry-3	Tennessee Valley A.	1,152	8- 8-76	3- 1-77	205	70.7
4. Cooper Station	Nebraska Pub. Pow. D.	836	2-21-74	7- 1-74	131	55.7
5. Fitzpatrick	Power Authority of N.Y.	883	11-17-74	7-28-75	253	45.4
6. Ft. Calhoun	Omaha Public Pow. Dist.	502	8- 6-73	9-26-73	51	52.1
7. Indian Point-3	Power Authority of N.Y.	1,013	4- 6-76	8-30-76	146	79.4
8. Rancho Seco	Sacramento Munic. Ut. D.	963	9-16-74	4-17-75	213	24.7

Sources: See table A-4.

Bibliography

American Nuclear Society. *Proceedings: Reliable Nuclear Power Today.* LaGrange Park, Ill., April 1978.

Arthur D. Little, Inc./S.M. Stoller Corp. "Economic Comparison of Base-Load Generation Alternatives for New England Electric." Cambridge, Mass., 1975.

Barth, Jacques. *40 Ans d'Energie Nucleaire dan le Monde.* Palaiseau, France: SOFEDIR, 1981.

Bown, Robert, and Snyder, Artha. *Total Energy, Electric Energy, and Nuclear Power Projections.* ERDA mimeograph, February 1975.

Electric Power Research Institute. *EPRI Journal,* monthly. Palo Alto, Calif.

International Atomic Energy Agency. *Operating Experience with Nuclear Power Plants in Member States,* annual reports.

Komonoff, Charles. *Power Plant Cost Escalation.* New York: Komonoff Energy Associates, 1981.

MITRE. "Analysis of Benefits Associated with the Introduction of Advanced Generating Technologies: Description of Methodologies and Summary of Results." March 1977. Prepared for U.S. ERDA.

Nuclear Energy Policy Study Group. *Nuclear Power Issues and Choices.* Sponsored by the Ford Foundation, administered by the MITRE Corporation. Cambridge, Mass.: Ballinger, 1977.

Nuclear Information and Resource Service. *Groundswell,* bimonthly.

Perry, Robert, et al. "Development and Commercialization of the Light Water Reactor, 1946–1976." Santa Monica, Calif.: Rand Corporation. 1977.

Shapiro, Fred C. *Radwaste.* New York: Random House, 1981.

Stobaugh, Robert, and Yergin, Daniel, eds. *Energy Future: Report of the Energy Project at the Harvard Business School.* New York: Random House, 1979.

TRW, "Electric Utilities Study: An Assessment of New Technologies from a Utility Viewpoint—Final Report." McLean, Va., November 1976. Prepared for the U.S. ERDA.

United Kingdom Nuclear Power Advisory Board. *Choice of Thermal Reactor Systems.* London: Her Majesty's Stationary Office, 1974.

——— House of Commons, Select Committee on Science and Technology. *The SGHWR Programme.* London: Her Majesty's Stationary Office, 1976.

U.S. Atomic Energy Commission. *The Nuclear Industry, Annual Review.* Washington, D.C.: Government Printing Office, various dates.

U.S. Department of Energy, Energy Information Administration. *Annual Report to Congress,* for 1981, vol. 3, "Energy Projections"; for 1980, vol. 3, "Forecasts." DOE/EIA-0173. Washington, D.C.: Government Printing Office.

——. "Commercial Nuclear and Uranium Market Forecasts for the United States and the World Outside Communist Areas." DOE/EIA-0184/24. Washington, D.C.: Government Printing Office, January 1980.

—— Energy Information Agency. *Nuclear Power Regulation, Energy Policy Study,* vol. 10, DOE/EIA-0201/10. Washington, D.C.: Government Printing Office, May 1980.

——. *Monthly Energy Review.* Washington, D.C.: Government Printing Office, various dates.

——. "Nuclear Power Program Information and Data," quarterly. *UPDATE.* Washington, D.C.

——. *U.S. Central Station Nuclear Electric Generating Units: Significant Milestones.* DOE/NE-0030/3(81). Washington, D.C., 1981.

——. *U.S. Commercial Nuclear Power: Historical Perspective, Current Status, and Outlook.* DOE/EIA-0315. Washington, D.C.: Government Printing Office, March 1982.

U.S. ERDA. "Comparing New Technologies for the Electric Utilities, Draft Final Report." ERDA 76-141, December 9, 1976.

U.S. Federal Power Commission/Federal Energy Regulatory Commission. *Steam-Electric Power Plant Construction Cost and Annual Production Expenses* (cited also as "Statistics of Steam-Electric Plants"). Washington, D.C.: Government Printing Office, various dates.

U.S. House of Representatives Committee on Government Operations. *Hearings, Nuclear Power Costs,* September 1977. Also *Report,* 1978.

U.S. Nuclear Regulatory Commission. "Coal and Nuclear: A Comparison of the Cost of Generating Baseload Electricity by Region." NUREG-0480. Washington, D.C., December 1978.

——. *Draft Environmental Statement, New England Power Units 1 and 2.* NUREG-0529. Washington, D.C., May 1979.

——. *Draft Generic Environmental Impact Statement on Decommissioning of Nuclear Facilities.* NUREG-0586. Washington, D.C., January 1981.

——. *Nuclear Power Plant Operating Experience,* annual. Washington, D.C.

——. *Operating Units Status Report,* monthly. Washington, D.C. (Often cited in the industry as "Gray Book.")

Wilson, Carroll. *Coal—Bridge to the Future.* Cambridge, Mass.: Ballinger, 1980.

Index

Adjustments to case studies:
calculation of 101–155; general
description of, xvi, xvii, 51–54;
general rationale for, 47–100
ADL. *See* Arthur D. Little
Advanced gas-cooled reactor (AGR),
12–15
AEA. *See* Atomic Energy Agency
(U.K.)
AEC. *See* Atomic Energy Commission
AEP. *See* American Electric Power
Company
Air temperature, 68–69
ALARA: As Low As is Reasonably
Achieveable, 93–100
American Electric Power Company
(AEP), 9, 57, 66, 74–76, 97
Argentina, 20–21
Arkansas Power and Light, 57
Arthur D. Little (ADL), 66–67, 72–73.
See also S.M. Stoller Corporation
Atomic Energy Agency (AEA, U.K.),
13–15
Atomic Energy Commission (AEC),
64, 68, 91; case study, xv, 47–54,
73, 107–118; detailed calculations
for base case lifetime costs,
109–113; detailed calculations for
initial commerciality costs, 116–118;
detailed calculations for sensitivity
analysis on lifetime costs, 114–116;
development of lifetime costs for
case study, 107; general rationale
for case study adjustments, 55–100;
summary of case study results,
103–106, 107–109
Atucha, Argentina, 20–21
Authorized power, 69
Automatic shutdown (scram), 19, 86
Automation. *See* controls
Availability of units, 17, 22, 69, 70,
85–87; TVA improvements on, 55;
See also Predictability of unit
availability; Yoyo factor; Operating
availability factor; Equivalent
availability factor

Baden, Germany, 17, 66
Base case: defined, 52–53
Battelle Pacific Northwest Laboratories
(PNL), 94
BCEGB. *See* British Central Electricity
Generating Board
Bechtel Corporation, xv, 4–6, 90
Bechtel theorem, xvii, 4–6, 90
Big Rock Point nuclear power plant,
96
Boiling-water reactor (BWR), 92, 96
Bown, Robert, 71
British Central Electricity Generating
Board (BCEGB), 13–15, 66
Broom, Trevor, 77
Brown-Boveri, 15
Browns Ferry nuclear power plant, 70,
87
Btu per pound of coal, xvi, 51, 53, 79;
See also individual case studies
Bull Run coal-fired power plant, 8
BWR. *See* Boiling-water reactor

Canfield, Monte, 98
California Energy Resources,
Conservation and Development
Commission, 99
Capacity Factor (CF), 18, 82, 89, 94,
157; adjustment to case studies, xvi,
51, 53, 55, 63, 67–76; assumptions
for ERDA case study, 50–51, 71;
definitions and concepts, 67–70;
Germany v. U.S., 15–17; nuclear
industry average, xvii, 9, 74, 85. *See
also* Senescence: Learning; Hydro
analog; Reliability, related costs;
individual case studies

175

About the Authors

Richard Hellman is a professor of economics at the University of Rhode Island and has served as director of the Research Center in Business and Economics. During the sabbatical year 1976–1977, he studied nuclear-power economics in Europe. Professor Hellman has served as director of economic planning and research and deputy assistant administrator at the U.S. Small Business Administration in Washington. His other government positions include economist in the Department of Commerce, chairman of the Government–Industry Task Force on Industrial Production after a Nuclear Attack, director of a study of the location of the first commercial nuclear power plant, assistant to the chairman and economist to the Bureau of Power at the Federal Power Commission, and economist in the U.S. Bureau of Labor Statistics. He served as consultant to the Federal Energy Administration and has been an expert witness for state governments in electric-utility-rate cases. He was a member of the Regulatory Task Force of the National Advisory Committee to the Federal Power Commission's National Gas Survey and a member of its drafting committee.

Professor Hellman's publications include a book titled *Government Competition in the Electric Utility Industry of the United States* (1972). He has been a Fellow at Brookings Institution and a Fulbright scholar in Paris. Professor Hellman received the B.A. and Ph.D. in economics from Columbia University.

Caroline J.C. Hellman is a research assistant professor in energy economics in the University Energy Center at the University of Rhode Island (URI). Before coming to URI, she was a consultant in the Energy Management and Strategy Group at Temple, Barker and Sloane, Inc. (TBS), in Lexington, Massachusetts and has served as an economist in the Office of Economics at the Federal Power Commission. Professor Hellman has assisted state and federal governments with policies pertaining to electric-utility-rate design, load management for electricity, alternative energy resources, and the National Energy Act. She coauthored a paper for the Federal Energy Administration titled "Risk Analysis and Strategies to Maximize Private Risk Taking in Substitute Gas Projects," and, for the Electric Utility Rate Design Study, wrote "Topic Paper 5: Customer Acceptance of Time-Differentiated Rates and Load Controls." Professor Hellman received the M.A. in economics and philosophy from Oxford University and the B.A. in economics from Cornell University.